Mathematical Modelling

MATHEMATICAL MODELLING:
Theory and Applications

VOLUME 6

This series is aimed at publishing work dealing with the definition, development and application of fundamental theory and methodology, computational and algorithmic implementations and comprehensive empirical studies in mathematical modelling. Work on new mathematics inspired by the construction of mathematical models, combining theory and experiment and furthering the understanding of the systems being modelled are particularly welcomed.

Manuscripts to be considered for publication lie within the following, non-exhaustive list of areas: mathematical modelling in engineering, industrial mathematics, control theory, operations research, decision theory, economic modelling, mathematical programmering, mathematical system theory, geophysical sciences, climate modelling, environmental processes, mathematical modelling in psychology, political science, sociology and behavioural sciences, mathematical biology, mathematical ecology, image processing, computer vision, artificial intelligence, fuzzy systems, and approximate reasoning, genetic algorithms, neural networks, expert systems, pattern recognition, clustering, chaos and fractals.

Original monographs, comprehensive surveys as well as edited collections will be considered for publication.

The titles published in this series are listed at the end of this volume.

Mathematical Modelling

Concepts and Case Studies

by

J. Caldwell
City University of Hong Kong,
Kowloon, Hong Kong

and

Y. M. Ram
Louisiana State University,
Baton Rouge, U.S.A.

KLUWER ACADEMIC PUBLISHERS
DORDRECHT / BOSTON / LONDON

A C.I.P. Catalogue record for this book is available from the Library of Congress.

ISBN 978-90-481-5263-6

Published by Kluwer Academic Publishers,
P.O. Box 17, 3300 AA Dordrecht, The Netherlands.

Sold and distributed in North, Central and South America
by Kluwer Academic Publishers,
101 Philip Drive, Norwell, MA 02061, U.S.A.

In all other countries, sold and distributed
by Kluwer Academic Publishers,
P.O. Box 322, 3300 AH Dordrecht, The Netherlands.

Printed on acid-free paper

CONTENTS

PREFACE

Over the past decade there has been an increasing demand for suitable material in the area of mathematical modelling as applied to science and engineering. There has been a constant movement in the emphasis from developing proficiency in purely mathematical techniques to an approach which caters for industrial and scientific applications in emerging new technologies. In this textbook we have attempted to present the important fundamental concepts of mathematical modelling and to demonstrate their use in solving certain scientific and engineering problems.

This text, which serves as a general introduction to the area of mathematical modelling, is aimed at advanced undergraduate students in mathematics or closely related disciplines, e.g., students who have some prerequisite knowledge such as one-variable calculus, linear algebra and ordinary differential equations. Some prior knowledge of computer programming would be useful but is not considered essential. The text also contains some more challenging material which could prove attractive to graduate students in engineering or science who are involved in mathematical modelling.

In preparing the text we have tried to use our experience of teaching mathematical modelling to undergraduate students in a wide range of areas including mathematics and computer science and disciplines in engineering and science. An important aspect of the text is the use made of scientific computer software packages such as MAPLE for symbolic algebraic manipulations and MATLAB for numerical simulation.

The book is divided into two main parts. Part I deals with continuous and discrete modelling. The starting point here is the background and fundamental concepts of mathematical modelling. In the formulation of mathematical models (Chapter 1) it is important to understand the mathematical modelling process, the input-output principle, simple rate models, models involving recurrence relations (i.e., discrete models), optimization models and sensitivity analysis.

In Chapter 2 (Compartment Problems) we explore simple linear problems which involve a single compartment (e.g., a room or tank) and the time rate of change of some specified quantity (e.g., heat energy or volume of liquid) within that compartment as it interacts with its environment. The work is extended to deal with multi-compartment problems and mathematical models are developed for problems involving systems of tanks including pollution.

Chapter 3 deals with continuous time models in the important areas of dynamics and vibration. Elementary concepts and principles are introduced and applied in practice by considering several dynamic systems and their mathematical models. The response of single degree of freedom systems is considered and important aspects such as equilibrium, linearisation and stability are dealt with. This work is extended for the more advanced reader to consider both modelling and response of multi-degree-of-freedom systems.

For some applications, e.g., in digital system control, the dynamics of the system is sampled after equal time intervals and so the response of the system involves discrete functions as opposed to continuous functions. So Chapter 4 concentrates on discrete models and includes some essential background material on finite-difference approximations and analytical solutions for linear systems of difference equations. The concept of stability is discussed and examples of discrete time models of dynamic systems are included. This work culminates in the finite-difference model of an eigenvalue problem by considering the problem of an axially vibrating rod.

Chapter 5 deals with numerical techniques for model nonlinear partial differential equations (PDE's) and is aimed primarily at graduate students. The example chosen to illustrate a range of possible numerical techniques for solution of nonlinear PDE's is that of Burgers' equation. The reason for this choice is that Burgers' equation is probably one of the simplest nonlinear PDE's for which it is possible to obtain an exact solution. Also, depending on the magnitude of the various terms in the equation, it behaves as an elliptic, parabolic or hyperbolic PDE. Hence it is very suitable as a model equation for testing and comparing numerical techniques. Such numerical techniques include finite-difference and Fourier methods, the method of lines, finite element methods with both fixed and moving nodes and variational-iterative schemes. Reference is made to past work by Caldwell and other co-researchers where more detailed information can be obtained from the list of bibliography.

The second part of the text, Part II, consists of a number of realistic Case Studies which illustrate the use of the modelling process in the solution of continuous and discrete models. These models are chosen to illustrate the use of the concepts, principles and techniques discussed in Part I.

Case Study A involves the design of biotreatment systems to process the effluent from a steel works. The inspiration here comes from former colleagues Professor Mark Cross, Professor Alfredo Moscardini and Yang, Yiu Wah.

Case Study B involves a deterministic model in the theory of contagious disease. The inspiration here comes from former colleague Dr Colin Creasy and students Andrew Biersterfield and Tse, Hoi Yan.

Case Studies C and D deal with analysis and synthesis of two mechanisms, namely, the cam-follower and the slider-crank. The function of the cam-follower mechanism is to transform rotational motion to linear. The synthesis problem of determining the parametric equations of the contour of the disk cam required to achieve a prescribed follower motion is studied in Case Study C. This essentially geometrical geometry problem involves in real life applications evaluation of some numerical derivatives of first and second order. The phenomenon of catastrophic cancellation associated with numerical differentiation is involved and a least squares smoothing technique is applied to overcome this difficulty.

In Case Study D we determine the position, velocity and acceleration of a slider-crank mechanism, and then evaluate the force applied on the pins connecting the various links. This problem demonstrates analysis of a dynamic system and its application in mechanism design. Both Case Studies C and D are solved via the use of MATLAB.

Each Chapter of Part I ends with a section comprising a number of exercises to be undertaken by students using the text. Challenging problems are also included at the end of each Case Study. To help make the material as clear as possible a number of worked test examples have been included in appropriate parts of the text.

We would like to express our thanks to other authors, former and present modelling colleagues and final year undergraduate students (too numerous

XII

to mention by name) who have helped over the past years to give us some of the important modelling ideas contained in this textbook. Finally, we wish to thank Chiu, Chi Keung for his part in the skillful typing of the manuscript.

We hope you find this book both interesting and challenging!

Dr Jim Caldwell
Dr Yitshak Ram

Part I

CONTINUOUS & DISCRETE MODELLING

Chapter 1

FORMULATION OF MATHEMATICAL MODELS.

1.1 The Mathematical Modelling Process

The use of mathematics in solving real-world problems is often referred to as 'mathematical modelling'. We could define this as the process of describing a real-world situation using mathematics. However, it must be done in such a way that it helps in the solution of the given problem. The idea of 'modelling' should be emphasized in that it helps to remind us that we must pay attention to things other than mathematics. The following flowchart in Figure 1.1 showing the various stages of the mathematical modelling process is by now well accepted and has been published by many authors of mathematical modelling textbooks.

At stage 1 the modeller has to become familiar with the problem he is trying to solve and to be clear about his objectives. The next stage involves starting to build the model. Here the modeller must decide which features of reality are to be included in the model and which can be neglected. Stage 3 involves formulation of the mathematical problem. This is the crucial stage and often the most difficult one. It involves deciding on symbols for the variables previously chosen at stage 2. Also any information we have about the real situation must be translated into equations or other mathematical statements. It is possible to use a number of techniques to translate the original question (or objective) into a question (or objective) about these equations, e.g., how to recognize and formulate proportionality relations and linear relations between the variables; how to use the 'input-output principle', i.e.

$$Increase = Input - Output$$

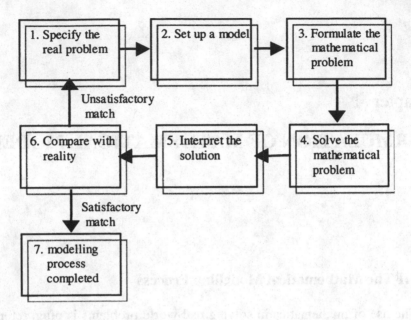

Figure 1.1 A flow-chart showing the various stages of the modelling process.

in formulating differential equations. (This important issue will be discussed in more detail in Section 1.2).

Stage 4 involves the process of solving the mathematical problem. Obviously skill and experience are important qualities here. Decisions need to be made on the method of solution (analytical or numerical) and whether to opt for a simple approximate method of solution or a more complicated accurate method. A detailed check on the mathematics is important.

Stage 5 involves the interpretation of the results of the model. One should consider whether the model behaves reasonably when changes are made to some of the conditions. The accuracy to be expected of the model and the domain of validity are other points to be considered here.

In comparing the model with reality at stage 6 we need to carry out the processes of validation, evaluation and iteration. If we feel that the model can be improved we have the opportunity of returning to stage 1 and going round the loop again. Mathematical modelling can be regarded as an iterative process where we start from a crude model and gradually refine it until it is good enough to solve the original problem.

1.2 Input-Output Principle

One of the fundamental principles which is frequently used in the formulation of the mathematical problem at stage 3 is the so-called 'input-output principle'. It is useful in suggesting relations between variables, namely,

$$\text{Increase} = \text{Input} - \text{Output} \tag{1.1}$$

This principle can easily be illustrated using population change models. The obvious reasons why the numbers in a population change are that individuals are born and individuals die. This suggests the equation

Increase in population = Births during the period − Deaths during the
period   (1.2)

One could ask the question: Does equation (1.2) cover all possible inputs and outputs to a population? The answer is 'no'. In order to take into account
(a) removals from the population by other forms of exploitation (which we can categorize as 'deaths');
(b) the movement of populations in space causing inputs and outputs to particular populations;
we should modify equation (1.2) as follows:

Increase in population = (Births + Immigration) − (Deaths + Emigration)
      (1.3)

This general equation for population change simplifies to equation (1.2) if there is no migration.

To proceed further, these ideas must be translated into mathematical relations. The input-output principle leads naturally to mathematical relations either in the form of recurrence relations or of differential equations.

Recurrence relations. First, we consider the formulation of the input-output principle in terms of recurrence relations.

Let y_k denote the size of a population k years after some specified initial time. Then the increase in population in year $k+1$ is $y_{k+1} - y_k$.

For a model which assumes no migration or exploitation, equation (1.3) becomes

$$y_{k+1} - y_k = (\text{Births in year } k+1) - (\text{Deaths in year } k+1) \qquad (1.4)$$

In order to formulate the mathematical problem (i.e., Stage 3 of the modeling process) we will make the following assumptions about births and deaths:

(a) The annual births are proportional to the population at the start of the year.

(b) The annual deaths are proportional to the population at the start of the year.

These are the simplest assumptions which seem reasonable. This then gives

$$\text{births in year } k+1 = by_k ,$$

and

$$\text{deaths in year } k+1 = dy_k ,$$

where b and d are constants. Therefore equation (1.4) becomes

$$y_{k+1} - y_k = (b-d)y_k . \qquad (1.5)$$

This is a first order difference equation, which can be rewritten as

$$y_{k+1} - (1+b-d)y_k = 0 .$$

The solution is

$$y_k = (1+b-d)^k y_0 \qquad (1.6)$$

with takes the form of an exponential model.

Differential equation model. Now we represent the population $y(t)$ as a continuously varying function of time t. As before we assume that

(a) there is no migration or exploitation;

(b) births and deaths in a short time period are proportional to the population at the beginning of the time period;

(c) births and deaths are proportional to the duration of the time period.

Now consider a short time period of duration h, starting at t. During this period

$$\text{Increase in population} = y(t+h) - y(t)$$
$$\text{Input to population} = \text{births} = By(t)h$$
$$\text{Output from population} = \text{deaths} = Dy(t)h$$

where B and t are constants.

The input-output principle gives

$$y(t+h) - y(t) = By(t)h - Dy(t)h$$
$$= (B-D)hy(t)$$

Rearranging we get

$$\frac{y(t+h) - y(t)}{h} = (B-D)y(t).$$

This equation holds for all small values of h, and so we can let h tend to zero. In the limit as $h \to 0$ we have

$$\frac{dy(t)}{dt} = (B-D)y(t) \tag{1.7}$$

In this way we have set up a differential equation model.

1.3 Rate Models

In application of mathematics we often require to maximize or minimize a certain quantity, e.g. in economics, we require to maximize profit, and in construction, we require to minimize the cost of materials. Problems of these types are generally referred to as extremum problems. In general, in an extremum problem we form a function $f(x)$ to represent the quantity to be maximized or minimized. We then obtain the derivative $f'(x)$ and solve $f'(x) = 0$ to obtain the solution of the extremum problem. The important point is that once the model $f(x)$ is obtained, it is then possible to apply calculus to solve the problem efficiently.

In this section we consider a range of rate models starting with examples described by the differential equation

$$y' = ky \qquad (1.8)$$

which relates the rate of change of y to the value of y. Equation (1.7) in the population model discussed in Section 1.2 fits into this category. In fact, in many cases the same mathematical model can be used to represent different problems. This has the advantage that, to obtain solutions to different problems, the model only requires to be solved once to formulate a general solution of the form

$$y = Ce^{kt}. \qquad (1.9)$$

We can therefore gain some insight from one problem represented by the model into other problems represented by the same model. There are many examples of this type of model including population growth rate, growth rate of a yeast culture in solution, compound interest and radioactive decay.

Test Example 1.1

Suppose a culture of bacteria B has a doubling time of 1.3 hours in a medium. The initial population is 5×10^6 per millilitre.
(a) Find the growth constant k.
(b) Find the population P as a function of time t.
(c) Find the time taken for the population to reach 10^8 per millilitre.

Solution

The rate model is $P' = kP$ with solution $P = Ce^{kt} = f(t)$, where k is the growth constant.

(a) Let D = doubling time

$$\therefore f(t+D) = 2f(t)$$
$$\therefore Ce^{k(t+D)} = 2Ce^{kt}$$
$$\therefore e^{kD} = 2$$

Hence $D = \ln 2 / k$ and so $1.3 = \ln 2 / k$ giving $k = 0.5332$.

(b) $P = Ce^{kt}$ But $P_0 = 5 \times 10^0 = C$

$$\therefore P = 5 \times 10^6 e^{0.5332t}$$

(c) $\ln P = \ln C + kt$

$$\therefore t = \frac{1}{k} \ln \left(\frac{P}{C} \right)$$

When $P = 10^8$ and $C = 5 \times 10^6$, we have

$$t = \frac{1}{0.5332} \ln 20 = 5.6184 \text{ hours}$$

Thus $t = 5$ hours 37 minutes to reach $P = 10^8 / \text{ml}$.

We will see that the population model (1.8) leads to unlimited exponential growth which is not realistic for large populations since a population will eventually level off due to environmental limitations. Thus the model (1.8) requires modification to obtain a model that levels off. We then progress to the model

$$y' = k(r - y) \tag{1.10}$$

which has the property that if y is close to r, then the slope y' is close to zero. In particular, the solution y levels off at r. This model (1.10) can be used as a rate model for

- drug buildup
- feedback control
- radioactive buildup

In each case the quantity $y = f(t)$ under consideration will converge to an equilibrium value r, i.e.,

$$\lim_{t \to \infty} f(t) = r.$$

This can be clearly seen from the solution to equation (1.10), namely,

$$y = r + Ce^{-kt}. \tag{1.11}$$

In particular, the difference $r - f(t) = r - y$ will converge to zero. For example, if an object with no internal heat source is placed in a room, then the temperature y of the object will converge to room temperature r as time increases. This explains why a cold glass of beer will warm up to room temperature whereas a hot cup of tea will cool down to room temperature.

Test Example 1.2

Suppose that for a certain drug the danger level for a certain part of the brain is $D = 0.0015$ grams. If the intake rate is 0.0005 grams/week and the removal rate constant k is 0.02, decide whether brain damage could occur. If so, estimate the time to reach danger level.

Solution

For the mathematical model, we assume that there is a constant rate d of drug intake and a removal rate ky. Hence the total rate of change of y is given by

$$y' = \text{total rate of change of } y$$
$$= \text{intake rate} - \text{removal rate}$$
$$= d - ky = k\left(\frac{d}{k} - y\right)$$

Letting $r = d/k$ gives us the model equation (1.10) which has solution (1.11).

In our case $d/k = 0.0005/0.02 = 0.025$. As this is greater than the danger level $D = 0.015$, brain damage could occur.

The time t for y to reach the danger level $D = 0.015$ is computed from equation (1.11). Clearly

$$\ln(y - r) = \ln C - kt$$

$$\therefore t = \frac{1}{k} \ln \frac{C}{y - r}$$

To find C, assume $y_0 = 0$. This gives $C = -r = -0.025$. Hence the time to reach $y = D = 0.015$ is given by

$$t = \frac{1}{0.02} \ln\left(\frac{-0.025}{0.015 - 0.025} \right)$$
$$= 50 \ln 2.5$$
$$= 45.8 \text{ weeks.}$$

This is a fairly long time to reach the danger level. Also, as the change from day to day is very slight, a person may not realize that a change is taking place until it is too late.

However, the model (1.10) is not appropriate as a population model since y' decreases as y increases for all values of y less than r. A population model should have the property that the rate of growth y' increases as the population y increases. This will happen while y is small enough so that limitations due to the environment are not included. It is important to note that it is possible to combine both the exponential growth property of equation (1.8) with the levelling off property of equation (1.10) into a single ordinary differential equation of the form

$$y' = ky(r - y) \tag{1.12}$$

This differential equation will provide a more realistic rate model for population growth. It can be used as a rate model for a variety of problems including spread of disease and autocatalytic reactions. The general solution of (1.12), namely,

$$y = \frac{r}{1 + Ce^{-krt}} \tag{1.13}$$

has exponential growth for small y and levels off for y near r.

Text Example 1.3

Let r denote the population that is susceptible to a certain disease. Let P be the population that has the disease. Devise a reasonable mathematical model for the spread of disease.

Solution

Clearly $r - P$ is the population that is susceptible to the disease but has not yet contracted the disease.

The disease will spread by contact between individuals. The more infected people P and the more susceptible people $r - P$, the more quickly the disease will spread.

Therefore it is reasonable to assume that the rate of change of the population that has the disease is proportional to the product $P(r-P)$.

Hence there is a positive constant k such that

$$P' = kP(r - P).$$ (1.14)

This has the form of equation (1.12) with solution

$$P = \frac{r}{1 + Ce^{-krt}}.$$ (1.15)

Test Example 1.4

Suppose population is measured in units of 1000, and the number of people in a town susceptible to a certain type of flu is 20,000. Assume the population P with the flu satisfies the model (1.14) with $k = 0.03$, where time is measured in weeks. Given that the initial population with the flu is 700, find

(a) P when $t = 3$ weeks.
(b) t and P' when $P = r/2$.
(c) t when $P = 8$.

Solution

We adopt the model $P' = kP(r - P)$ where $r = 20$. Since $k = 0.03$, we have

$$P' = 0.03P(20 - P).$$

Using (1.15) the solution is

$$P = \frac{20}{1 + Ce^{-0.6t}}$$

(a)   When $t = 0$, $P = 20/(1 + C)$.   But $P_0 = 0.7$,   and   so
$C = (20 - 0.7)/0.7 = 27.57$.

$$\therefore \text{ When } t = 3, P = \frac{20}{1 + 27.57 e^{-1.8}} = 1.131$$

Thus 1131 people have the flu after 3 weeks.

(b) From equation (1.15) we find that

$$t = \frac{1}{kr} \ln\left(\frac{CP}{r - P} \right). \tag{1.16}$$

Hence

$$t = \frac{1}{0.6} \ln\left(\frac{27.57 P}{20 - P} \right).$$

Thus when $P = r/2 = 10$, $t = (1/0.6) \ln 27.57 = 5.528$. This means that it takes approximately $5\frac{1}{2}$ weeks for 10,000 people to catch the flu.

Also when $P = r/2 = 10$, $P' = 0.03 P(20 - P) = 0.03(100) = 3$. This means that the flu is spreading at a maximum rate of 3000 cases per week after $5\frac{1}{2}$ weeks.

(c) If $P = 8$, then from (1.16) we have

$$t = \frac{1}{0.6} \ln\left(\frac{27.57 \times 8}{12} \right) = 4.852.$$

Thus there are 8000 cases of flu in just under 5 weeks.

Finally, this work can be extended to consider models which represent the growth process of two competing or interacting species whose survival depends upon their mutual cooperation, e.g. predator-prey interaction. This means that we can extend equation (1.12) to produce a rate model involving a pair of coupled first order ordinary differential equations. Using variables x and y to represent the numbers in each population we obtain a mathematical model of the form

$$\begin{aligned} x' &= k_1 x(r_1 - x) - \alpha_1 xy \\ y' &= k_2 y(r_2 - y) - \alpha_2 xy \end{aligned} \tag{1.17}$$

where the additional parameters α_1, α_2 are introduced to represent the effects of competition.

1.4 Models Involving Recurrence Relations

Recurrence relations (or difference equations) have been mentioned in Section 1.2 It is important to classify them in terms of order and linearity. A recurrence relation of k th order is one where the difference between the highest and lowest subscripts is k, e.g.,

$u_{k+1} = 3u_{k-1} + 5$ is 2nd order.

A linear recurrence relation is one that can be written as

$u_{k+1} = a_k u_k + b_k u_{k-1} + c_k u_{k-2} + \cdots + s_k,$

where the coefficients a_k, b_k, c_k, ..., s_k may depend on k but not on u. Otherwise it is said to be non-linear, e.g.,

$u_k = 2ku_{k-1} - u_{k-2}$ is linear,

but

$u_k = 4u_{k-1}^2 + 5u_{k-2} + ku_{k-3}$ is non-linear.

Now we consider some simple models involving recurrence relations which include modelling a mortgage and application to economic models. Such models are referred to as discrete models as opposed to the continuous models discussed in Section 1.3.

Test Example 1.5 (Repayment Mortgage)

A repayment mortgage is a system of debt repayment by fixed periodic instalments over an agreed interval of time. Between instalments the debt remaining accumulates compound interest at a fixed rate r. Assume the original amount borrowed is $\$B$ and that the percentage interest rate during the time intervals between repayments is I ($=100r$).

The periodic repayment $\$R$ will consist partly of interest due and partly the repayment of the amount $\$B$ borrowed. Provided R is sufficiently large, the debt will be paid off in a finite number of repayments.

(a) Assuming $\$P_k$ is the outstanding debt after k repayments of $\$R$ have been made, develop a mathematical model in terms of difference equations.

(b) Solve this model and deduce that

$$P_k = \frac{R}{r} + (1+r)^k \left(B - \frac{R}{r} \right), \quad k = 0, 1, 2, \ldots$$

(c) If the debt is to be fully paid off in N equal repayments of $\$R$ show that

$$R = \frac{rB}{1 - (1+r)^{-N}}.$$

(d) Find the number of instalments required to pay off half of the original debt.

Solution

(a) Let $\$P_k$ = amount outstanding after k repayments have been made. The recurrence relation representing compound interest is given by

$$A_{k+1} = (1+r)A_k,$$

where A_0 = amount at the start, r = rate of interest ($I/100$). Hence

$$P_{k+1} = (1+r)P_k - R, \quad (k \geq 0)$$

where $P_0 = B$ and $r = I/100$. This is the important difference equation to be solved.

(b)

$$P_{k+1} - (1+r)P_k = -R, \tag{1.18}$$

As in the solution of differential equations, let the general solution be

$$P_k = P_k^c + P_k^p, \tag{1.19}$$

where P_k^c = complementary function (C.F.), P_k^p = particular integral (P.I.).

C.F. Let $P_k = \beta^k$. Substitute into the homogenous equation

$$P_{k+1} - (1+r)P_k = 0$$

to give

$$\beta^{k+1} - (1+r)\beta^k = 0$$
$$\therefore \beta - (1+r) = 0$$
$$\therefore \beta = (1+r)$$
$$\therefore P_k^c = c_1 (1+r)^k$$

where c_1 is an arbitrary constant.

P.I. Try $P_k^p = c_2$. Substitute into equation (1.18) to give

$$c_2 - (1+r)c_2 = -R$$
$$\therefore c_2 = \frac{R}{r}$$
$$\therefore P_k^p = \frac{R}{r}$$

Hence using (1.19) $P_k = c_1(1+r)^k + R/r$.

Now we can find c_1 using the initial condition that $P_0 = B$. This gives

$$B = c_1 + \frac{R}{r} \quad \text{and so} \quad c_1 = B - \frac{R}{r}.$$

Therefore

$$P_k = \frac{R}{r} + (1+r)^k \left(B - \frac{R}{r} \right), \quad k = 0, 1, 2, \dots \tag{1.20}$$

as required.

If the debt is to be fully paid off in N equal instalments of $\$R$ then the amount owing after the last repayment will be zero, i.e.,

$$P_N = 0.$$

To find the fixed periodic repayment R in this case we use

$$\frac{R}{r} + (1+r)^N \left(B - \frac{R}{r} \right) = 0$$

$$\therefore \frac{R}{r} \left\{ 1 - (1+r)^N \right\} = -B(1+r)^N$$

$$\therefore R = \frac{-rB(1+r)^N}{1 - (1+r)^N} = \frac{rB}{1 - (1+r)^{-N}}.$$

We now require to find the number of instalments N to pay off half of the original debt which is $B/2$. Hence using (1.20)

$$\frac{B}{2} = \frac{R}{r} + (1+r)^N \left(B - \frac{R}{r} \right)$$

$$\therefore (1+r)^N = \left\{ \left(\frac{B}{2} - \frac{R}{r} \right) \bigg/ \left(B - \frac{R}{r} \right) \right\}$$

$$\therefore N = \frac{\ln \left\{ (B/2 - R/r) / (B - R/r) \right\}}{\ln(1+r)}$$

In simple models from Economics difference equations can be used to express the dependence of the value of a variable on its value in the preceding period. The most popular Economics models include the Harrod model, general cobweb model, consumption model, income-consumption-investment model, etc.

The example below demonstrates the use of the cobweb model in the solution of a supply/demand problem.

Test Example 1.6 (Model from Economics)

Suppose supply and demand can be studied using

$$q_k = 2 + \frac{1}{2} p_{k-1} \quad \text{(supply)}$$

$$q_k = 10 - p_k \quad \text{(demand)}$$

where the price p and quantity q are both functions of time k. We are interested in the behaviour of the market in the long term. Develop a mathematical model using difference equations and solve.

Solution

For equilibrium, supply = demand. Hence

$$2 + \frac{1}{2} p_{k-1} = 10 - p_k$$

$$p_k + \frac{1}{2} p_{k-1} = 8 \qquad\qquad (1.21)$$

As in Test Example 1.5, $p_k = $ C.F. + P.I.

C.F. Try

$$p_k = \beta^k \quad \text{in} \quad p_k + \frac{1}{2} p_{k-1} = 0.$$

This gives

$$\beta^k + \frac{1}{2} \beta^{k-1} = 0$$

$$\therefore \beta = -\frac{1}{2}$$

C.F. is $c_1(-1/2)^k$.

P.I. Try $p_k = c_2$ in equation (1.21). This gives

$$c_2 + \frac{1}{2} c_2 = 8$$

$$c_2 = \frac{16}{3}$$

Hence the general solution of (1.21) is

$$p_k = c_1 \left(-\frac{1}{2}\right)^k + \frac{16}{3}.$$

At $t = 0$ let $p_k = p_0$ and so

$$p_0 = c_1 + \frac{16}{3}$$

$$\therefore c_1 = p_0 - \frac{16}{3}$$

$$\therefore p_k = \left(p_0 - \frac{16}{3}\right)\left(-\frac{1}{2}\right)^k + \frac{16}{3}$$

Also

$$q_k = 10 - p_k$$

$$= -\left(p_0 - \frac{16}{3}\right)\left(-\frac{1}{2}\right)^k + \frac{14}{3}.$$

Since

$$\lim_{k \to \infty} p_k = \frac{16}{3} \quad \text{and} \quad \lim_{k \to \infty} q_k = \frac{14}{3}$$

the market would oscillate and converge to the equilibrium point $(p^*, q^*) = (16/3, 14/3)$ as shown in Figure 1.2.

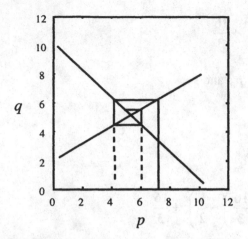

Figure 1.2 Cobweb phenomenon (cycle)

Test Exercise 1.7 (Modelling a Mortgage)

Consider the case of a $200,000 loan which is paid back over 15 years when the interest rate is 15% per annum. At the beginning of each year the 15% interest is added to the amount outstanding. During the following 12 months the repayments are made at monthly intervals, e.g. the interest charged in the first year is 15% of $200,000 which is $30,000. If M is the monthly payment, the amount owed at the end of the first year is

$$300,000 + 30,000 - 12M \ .$$

Interest is then charged in the second year based on this amount, and so on. Develop a mathematical model and solve to find the monthly repayment M.

Solution

In the modelling process we will keep the problem as simple as possible by not taking into account any extra payments, such as house insurance and payment of fees. We simply have a mortgage which has to be repaid over a 15 year period.

The basic equation for the amount owed next year is

$$\begin{bmatrix} \text{Amount owed} \\ \text{next year} \end{bmatrix} = \begin{bmatrix} \text{Amount owed} \\ \text{this year} \end{bmatrix} + [\text{Interest}] - [\text{Repayments}] \quad (1.22)$$

Let u_k represent the amount owed after k years. Then our equation (1.22) can be written as a recurrence relation:

$$u_{k+1} = u_k + 0.15 u_k - 12M,$$

i.e.

$$u_{k+1} = 1.15 u_k - 12M. \quad (1.23)$$

As explained before the solution of this first order difference equation is given by

$$u_k = \text{C.F.} + \text{P.I.}$$

C.F. Try $u_k = \beta^k$ in $u_{k+1} - 1.15 u_k = 0$
to obtain

$$\beta^{k+1} - 1.15 \beta^k = 0$$
$$\therefore \beta = 1.15$$
$$\therefore \text{C.F. is } c_1 (1.15)^k.$$

P.I. Try $u_k = c_2$ in equation (1.23). This gives

$$c_2 - 1.15 c_2 = -12M$$
$$\therefore c_2 = \frac{12M}{0.15} = 80M.$$

Hence the general solution of (1.23) is

$$u_k = c_1 (1.15)^k + 80M$$

where c_1 is an arbitrary constant.

As this solution contains two unknowns c_1 and M, we need two conditions to evaluate them. However, we know that the initial mortgage is $200,000, i.e., $u_0 = 200,000$. Also the mortgage is repaid in 15 years, i.e.,

$u_{15} = 0$.

Hence we obtain two simultaneous equations:

$$200,000 = c_1 + 80M \qquad\qquad (1.24)$$

$$0 = c_1 (1.15)15 + 80M \qquad\qquad (1.25)$$

Subtracting gives $200,000 = c_1 \{1 - (1.15)15\}$ and so

$$c_1 = -2.8022.73686\ldots .$$

Substituting back into equation (1.24) gives

$$M = \frac{200,000 - c_1}{80} = 2850.28421\ldots .$$

The monthly repayment figure M is not an amount we can pay exactly. The best we can do is to pay \$2850 per month. Using $M = 2850$ we obtain the difference equation

$$u_{k+1} = 1.15u_k - 34,200$$

which could be used to determine the amount owed each year. Rounding off each value of u_k, as it is calculated, to the nearest dollar, we obtain the values in Table 1.1.

k	u_k	k	u_k
0	200,000	8	142,347
1	195,800	9	129,499
2	190,970	10	114,724
3	185,416	11	97,733
4	179,028	12	78,193
5	171,682	13	55,722
6	163,234	14	29,880
7	153,519	15	162

Table 1.1 Table showing the amount owed u_k after k years.

The under-payment of \$162 could easily be resolved by adjusting the payments in the final year.

1.5 Optimization Models

Problems in optimization are the most common applications of mathematics. Single variable optimization problems are sometimes referred to as maximum-minimum problems and only require elementary calculus. Optimization models are designed to determine the values of the control variables which lead to the optimal outcome given the problem constraints. The task of locating maximum and minimum points can be quite difficult even in simple single variable optimization problems. In fact, in real problems the computation of the derivative can often be complicated even when the functions involved are differentiable everywhere. In most cases the equation $f'(x) = 0$ cannot be solved analytically and so numerical methods are required.

Of course, multivariable optimization problems involve multivariable calculus and the simplest type involve finding the maximum or minimum of a differentiable function of several variables over a defined region. Further complications arise when the region over which we optimize becomes more complex. For real problems more complicated models arise where we have to impose constraints on the independent variables. The solution may require the use of Lagrange multipliers.

The simplest type of multivariable constrained optimization problem is one where both the objective function and the constraint functions are linear. Computational methods for such problems give rise to the area of Linear Programming and usually require the application of relevant software packages.

In this section we restrict our attention to the single variable case to demonstrate the formulation and solution of a typical optimization model.

Test Example 1.8

A car manufacturer makes a profit of $2000 on the sale of a certain model of sportscar. It is estimated that for every $100 of rebate, sales increase by 12%. What amount of rebate will maximize profit?

Solution

First we draw up a list of the variables.

Variables: r = rebate (\$), c = number of cars sold, p = profit (\$)

We will assume that the number of cars sold without rebate is $r = c_0$ (a constant). To get a feel for the problem we can do some numerical experiments from which we can establish relationships between the variables, i.e., the assumptions to be made.

Rebate r (\$)	Cars sold c	Profit p (\$)	Profit increase	%Profit increase
0	c_0	$2000\,c_0$	–	–
100	$1.12\,c_0$	$2128\,c_0$	$128\,c_0$	6.4
200	$1.24\,c_0$	$2232\,c_0$	$232\,c_0$	11.6
300	$1.36\,c_0$	$2312\,c_0$	$312\,c_0$	15.6
400	$1.48\,c_0$	$2368\,c_0$	$368\,c_0$	18.4
500	$1.60\,c_0$	$2400\,c_0$	$400\,c_0$	20.0
600	$1.72\,c_0$	$2408\,c_0$	$408\,c_0$	20.4
700	$1.84\,c_0$	$2392\,c_0$	$392\,c_0$	19.6
800	$1.96\,c_0$	$2352\,c_0$	$352\,c_0$	17.6
900	$2.08\,c_0$	$2288\,c_0$	$288\,c_0$	14.4
1000	$2.20\,c_0$	$2200\,c_0$	$220\,c_0$	11.0

Table 1.2 Calculation of percentage profit increase v. rebate r.

From Table 1.2 we can list the assumptions about the variables if we assume the number of cars sold without any rebate is v_0 (a constant).

Assumptions:

c_0 = number of cars sold without any rebate (constant).

$c \geq 0$

$0 \leq t \leq 2000$

$$c = c_0\left(1 + 0.12\left(\frac{r}{100}\right)\right)$$

$p = (2000 - r)c$

Now we are in a position to state the objective which is clearly to maximize the profit p. From the assumptions listed above we can easily formulate the mathematical model which is to maximize

$$p = f(r) = (2000 - r)c_0\left(1 + 0.12\left(\frac{r}{100}\right)\right) \tag{1.26}$$

Our goal is to maximize $f(r)$ over the interval [0, 2000]. For a maximum, $f'(r) = 0$ and so

$$c_0\{0.0012(2000 - r) - (1 + 0.0012r)\} = 0$$

Hence $0.0024r = 1.4$ giving

$$r = \frac{1.4}{0.0024} = \frac{1750}{3} = 583.\dot{3} \ .$$

Since $f''(r) = -0.0024c_0 < 0$ we have a maximum. At $r = 1750/3$,

$$\text{maximum } f(r) = c_0\left(\frac{4250}{3}\right)(1 + 0.7) = 2408.\dot{3} \ c_0 \ .$$

Since the graph of $f(r)$ is a parabola we know this is the global maximum.

The percentage increase in profit will be

$$\frac{(2408.\dot{3} - 2000)}{2000} \times 100\% = 20.41\dot{6}\% \ .$$

So according to this model the optimum policy is to offer a rebate of around \$583, which should result in a profit increase of around 20%.

It is interesting to note that this result is backed up by the numerical experiments carried out in Table 1.2. More precise information could be obtained by drawing a graph of rebate r v. percentage profit increase. The maximum point of the parabola produced would lie at $(583.3, 20.416)$.

1.6 Sensitivity Analysis

The mathematical modelling process begins by making some assumptions about the variables. In most problems we cannot always be certain that all of the assumptions will be exactly valid. This means that we must consider how sensitive our conclusions are to each of the assumptions made. This kind of sensitivity analysis is an important part of the mathematical modelling process.

To illustrate this point we refer back to the car sales problem in Test Example 1.8. There we made the assumption that for every \$100 of rebate, sales increase by 12%. However, it is very unlikely that there would be a constant percentage increase in sales for every \$100 of rebate. Also it would be difficult to give this percentage increase to any great degree of certainty. Let R denote the rate of sales increase/dollar. In our example $R = 0.0012$, but now suppose that the actual value of R is different. We can get some idea of the sensitivity of our answer to the value of R by repeating the solution procedure for several different values of R. In fact, Table 1.3 shows the results of solving the problem for a few selected values of R around $R = 0.0012$. Clearly the amount of rebate is very sensitive to the parameter R.

R	r (\$)
0.0008	375.00
0.0010	500.00
0.0012	583.33
0.0014	642.86
0.0016	687.50

Table 1.3 Sensitivity of the rebate r (\$) to the rate R of sales increase per dollar.

A more systematic method for measuring this sensitivity would be to treat R as an unknown parameter and follow the same steps as before. This means that we take

$$p = f(r) = c_0 (2000 - r)(1 + Rr) \tag{1.27}$$

and proceed as before.

Then $f'(r) = -c_0 (1 + Rr) + c_0 R(2000 - r)$ so that $f'(r) = 0$ at the point

$$r = \frac{2000R - 1}{2R} . \tag{1.28}$$

It is more useful to interpret sensitivity data in terms of relative change or percentage change rather than in absolute terms. For example, a 17% increase in R leads to a 10% increase in r. If r changes by an amount Δr, the relative change in r is $\Delta r / r$, and the percentage change in r is

$100 \Delta r / r$. If R changes by ΔR, resulting in the change Δr in r, then the ratio between the relative change is $\Delta r / r$ divided by $\Delta R / R$. Letting $\Delta R \to 0$ and using the definition of derivative gives

$$\frac{\Delta r / r}{\Delta R / R} \to \frac{dr}{dR} \cdot \frac{R}{r} \,.$$

This limiting quantity is called the sensitivity of r to R, and we denote it by $S(r, R)$.

In the car sales problem we have

$$\frac{dr}{dR} = \frac{1}{2R^2} = 347{,}222$$

at the point $R = 0.0012$. Thus

$$S(r, R) = \frac{dr}{dR} \cdot \frac{R}{r} = 347{,}222 \times \frac{0.0012}{583.3} = 0.714 \tag{1.29}$$

This means that if R increases by 10%, then r increases by 7%. So, in real terms, if the rebate is 10% more effective than we thought, then the optimal rebate will be 7% bigger.

Test Example 1.9

In the car sales problem of Test Example 1.8 compute the sensitivity of the resulting profit to the 12% assumption. Suppose now that rebates actually generate only a 10% increase in sales per $100. What is the effect?

Solution

We generalize the model in equation (1.26) resulting from Test Example 1.8 and let

$$p = f(r) = c_0 (2000 - r)(1 + Rr) \tag{1.30}$$

where currently $R = 0.0012$.

$$\therefore \; \frac{dp}{dR} = c_0 r (2000 - r)$$

Hence the sensitivity of the resulting profit p to R is given by

$$S(p,R) = \frac{dp}{dR} \cdot \frac{R}{p}$$

$$= c_0 r(2000 - r) \cdot \frac{R}{c_0(2000 - r)(1 + Rr)} \qquad (1.31)$$

$$= \frac{Rr}{1 + Rr}.$$

Setting $R = 0.0012$, $r = 583.3$ gives

$$S(p,R) = \frac{0.0012(583.\dot{3})}{1 + (0.0012)(583.\dot{3})} = \frac{0.7}{1.7} = 0.412 . \qquad (1.32)$$

So if the rebate is 10% more effective than we thought, then our optimal profit will be 4% greater.

Using the result in equation (1.29), if R decreases by 17% to 0.0010 then we expect the optimal rebate to decrease by $(0.7)17\% = 12\%$ to around \$513. Using the result in (1.32) we would expect profits to go down by $(0.41)17\% = 7\%$ to around $\$2240c_0$ where c_0 represents the number of sales without any rebate.

It is important to note that similar results could be obtained by direct computation, i.e., solving the problem again using $R = 0.0010$. This would give

$$p = f(r) = c_0(2000 - r)(1 + 0.0010r) .$$

$\therefore f'(r) = 0$ leads to

$$c_0\{0.0010(2000 - r) - (1 + 0.0010r)\} = 0 .$$

Hence $0.0020r = 1$ giving $r = 500$, i.e., a 14.3% decrease in the optimal rebate (to \$500).

At $r = 500$, the maximum $p = c_0(1500)(1.5) = 2250c_0$, i.e., a   7.6% decrease in profit (to $\$2250c_0$).

1.7 Exercises

1. Plutonium-239 is a common reactor product and ingredient of nuclear bombs. This element takes a very long time to decay. This is why a reactor explosion could be very serious since there would be long term after-effects. Plutonium-239 has an annual decay constant $\lambda = 285 \times 10^{-7}$.

  (a)   Find the amount P of plutonium-239 at time t if the initial amount is $P_0 = 10$ grams.

  (b)   How much remains after 500 years?

  (c)   Find t as a function of P.

2. Suppose interest is compounded continuously at an annual rate of 8% and the initial deposit is $1000.

  (a) Find the amount A in the account at time t.

  (b) Find A when $t = 10$ years.

  (c) Find the doubling time, i.e. the time for A to reach $2000.

  (d) Find t when $A = \$2400$.

3. Let T be the temperature of a cup of tea with initial $T_0 = 180°$. Suppose room temperature is $65°$. Assume T satisfies the rate model (1.10) with rate constant $k = 0.20$ and that time t is measured in hours.

  (a) Find the temperature T as a function of time t.

  (b) Find T when $t = 4$ minutes.

  (c) Find t when $T = 100°$

4. Show that the general solution of the rate model

$$y' = ky(r - y)$$

is given by

$$y = \frac{r}{1 + Ce^{-krt}}$$

where C is a constant.

  For the case when $k = 0.2$ and $r = 20$, find

  (a) the solution y if $y_0 = 3$.

  (b) an expression for t in terms of y.

  (c) t when $y = 8$.

  (d) the point of inflection in the graph of y v. t.

5. An ecological study of a lake indicates that the lake can support a

population of 4000 fish. The lake is initially stocked with 200 fish, and there are 800 fish after 6 months. Assume the population y satisfies the rate model (1.12), where y is measured in units of 100 and $r = 40$. Find
  (a) the rate constant k.
  (b) the population y when $t = 12$ months.
  (c) the time t for the population y to reach 38.
  (d) t and y' at the point of inflection of the y v. t graph.

6. In certain chemical reactions of a chemical A into a chemical B, the product B catalyzes the reaction. This is called autocatalytic reaction.

Let x denote the amount of B present, and let y denote the amount of A present. Since one molecule of A is converted into one molecule of B, then $x + y$ remains constant. Let the initial value of this sum be r, and hence

$x + y = r$

Show that a mathematical model can be devised of the form (1.12).

Now suppose A is converting to B in an autocatalytic reaction and the rate constant $k = 0.01$, where time is measured in minutes. The total amount of A and B is 10 litres and the initial amount of B is 1 litre. Find
  (a) the amount of B after 30 minutes.
  (b) t when the amount of B is 9 litres.

7. Find the monthly repayment for a $100,000 mortgage at 15% over
  (i) 15 years
  (ii) 20 years
  (iii) 25 years
  (iv) 30 years.
  Comment on your results.

8. If the repayment on a $100,000 mortgage is fixed at $1500 per month, how long will it take to pay off the mortgage if the interest rate is
  (i) 10%
  (ii) 12%
  (iii) 15%?

9. Consider the following model of a market:

$Q_{dt} = a - bP_t$ (demand)

$Q_{st} = -c + dP_t^*$ (supply)

$Q_{dt} = Q_{st}$

where a, b, c, d are positive constants and P_t^* is the expected price at period t so that

$$P_t^* = P_{t-1}^* + e\left(P_{t-1} - P_{t-1}^*\right) \qquad e \in (0,1).$$

(a) Formulate a mathematical model by finding a difference equation involving P_t only.
(b) Solve the model for the solution sequence of P_t.
(c) Show that the solution sequence of P_t converges when

$$1 - \frac{2}{e} < -\frac{d}{b}$$

and find the equilibrium solution.

10. Suppose p_k represents the price of corn at period k. The equation

$$S_{k+1} = S_0 - bp_k \qquad\qquad b, S_0 > 0$$

is the supply function of corn at period $k+1$. Also

$$d_{k+1} = d_0 - ap_{k+1} \qquad\qquad a, d_0 > 0$$

is the demand function at period $k+1$. Given the initial price p_0, find the general solution of p_k and investigate the limiting behaviour of the solution for the case $b > a$.

11. A pig weighing 300 pounds gains 8 pounds per day and costs 60 cents per day to keep. The market price for pigs is 60 cents per pound, but is falling 1 cent per day.
(a) Model as a single variable optimization problem to find the best time to sell the pig and the resultant maximum profit.
(b) Examine the sensitivity of the best time to sell to the growth rate of the pig.
(c) Examine the sensitivity of the resulting maximum profit to the rate at which the price for pigs is dropping.

12. The population growth rate (per year) of a certain species of fish is estimated to be $gx(1-x/k)$, where $g=0.10$ is the intrinsic growth rate, $k=500,000$ is the maximum sustainable population and x is the current population, now around 80,000. Also the number of fish harvested per year is about $0.00005E_x$, where E is the level of fishing effort (boat-days). Given a fixed level of effort, the fish population will eventually stabilize at the level where growth rate equals harvest rate.

(a) Model as a single variable optimization problem to find the level of effort required to maximize the sustained harvest rate.

(b) Examine the sensitivity to the intrinsic growth rate g. Consider both the optimum level of effort E and the resulting population level x.

(c) Examine the sensitivity to the maximum sustainable population. Consider both the optimal level of effort E and the resulting population level x.

Chapter 2

COMPARTMENT PROBLEMS.

2.1 Introduction

In this Chapter we explore simple linear problems which involve a single compartment (e.g., a room/tank) and the time rate of change of some specified quantity (e.g., heat energy/volume of liquid) within that compartment as it interacts with its environment. As an introductory example we consider a model for temperature change in a single room. This provides an example of a general class of problems called 'compartment' problems. They all involve changing amounts of some physical quantity, enclosed in a prescribed spatial region, and usually interacting with the exterior environment in some way.

Another similar problem which arises frequently is that involving a tank containing some liquid which may also contain some pollutant. It is possible to develop a mathematical model for a single tank system. This work can easily be extended to consider a number of inter-connected tanks (i.e., multiple-compartment-problems).

Such problems can be represented diagrammatically as a 'black box' R which models the prescribed spatial region (or multiple boxes R_1, R_2, R_3, ... for multi-compartment-problems), with $Q(t)$ the amount of the physical quantity contained in R.

Physical quantity enters $R \rightarrow \boxed{\begin{array}{c} Q(t) \\ R \end{array}} \rightarrow$ Physical quantity exits R

Irrespective of whether the quantity is matter (e.g., salt in a solution) or energy (e.g., heat, as measured by temperature), the appropriate conservation laws imply a basic model of the form.

$$\frac{dQ}{dt} = \left[\begin{matrix} \text{rate/unit time at which} \\ \text{the quantity enters } R \end{matrix}\right] - \left[\begin{matrix} \text{rate/unit time at which} \\ \text{the quantity exits } R \end{matrix}\right] \qquad (2.1)$$

2.2 Model for Temperature Change in a Single Room

Before formulating the mathematical model it is important to list possible factors affecting the temperature in a room. These include
- heat loss/gain through walls, floor and ceiling
- air-conditioning
- furnace heat
- radiant solar energy from sunshine
- heat released from refrigerators, etc.
- additional heat sources.

The heat transfer through boundary surfaces (see Figure 2.1) occurs according to Newton's law of cooling, namely

$$\frac{dT}{dt} = -k\{T(t) - T_\alpha(t)\} \qquad (2.2)$$

where $T(t)$ represents temperature, k depends on the specific type of surface (e.g., larger for windows than walls), and $T_\alpha(t)$ represents the temperature on the other side of the surface.

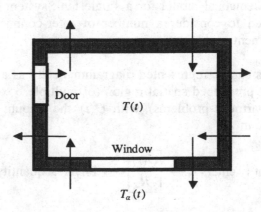

Figure 2.1 Temperature in a single room.

Thus a general model would have the form

$$\frac{dT}{dt} = -k\{T(t) - T_\alpha(t)\} + H(t) \tag{2.3}$$

where $H(t)$ refers to air-conditioning/furnace and other sources. We can say that the heating (or cooling) rate of the furnace (air-conditioning) will be approximately constant while it is operating. Other heat sources will depend on the specific cause.

This problem is quite a complex one but here we illustrate how it is possible to formulate a mathematical model by taking a relatively simple situation. In the modelling process it is important to list the simplifying assumptions. These are as follows:
(1) Consider a day which requires neither furnace heat nor air-conditioning.
(2) Consider a room on the main floor of a multi-storey building so that no heat transfer occurs through walls, floor or ceiling except for the window-wall to the exterior.
(3) Assume this window will also allow radiant energy gain from sunlight.
(4) Assume full sunlight falls on the window wall over the time period 9.00 a.m. to 12.00 noon.

From the above assumptions it is possible for us to make some reasonable approximations which will help in the mathematical formulation. For example, while the radiant energy gained actually varies both with the sun's compass direction and its altitude, a reasonable approximation can be obtained by averaging over the time period. Hence we assume a constant energy $H(t) = H_c$. Finally, the ambient exterior temperature $T_\alpha(t)$ will change significantly during the three hours time period of the model. However, we assume a linear increase

$$T_\alpha(t) = T_M + \alpha t$$

where T_M is the morning temperature at 9.00 a.m. and α is a constant.

Under the above assumptions, the model equation (2.2) becomes

$$\frac{dT}{dt} = -k\{T(t) - (T_M + \alpha t)\} + H_c, \quad T(0) = T_0 \tag{2.4}$$

where T_0 is the room temperature at 9.00 a.m., at which point we have assigned time $t = 0$. This simplifies to

$$\frac{dT}{dt} + kT = kT_M + k\alpha t + H_c, \quad T(0) = T_0 . \tag{2.5}$$

It is important to note that if we make the following two assumptions in the mathematical model:

(1) the exterior temperature $T_\alpha(t)$ is constant;
(2) the heating (or cooling) rate of the furnace (air-conditioning) is approximately constant while it is operating;

then it would be possible to solve equation (2.5) analytically by separation of variables. However, in the general case of equation (2.5), the mix of terms kT and $k\alpha t$ prevents solution by separation of variables. This means that we would have to resort to an approximate solution by a numerical method such as Euler's method.

Re-writing equation (2.4) in the form

$$\frac{dT}{dt} = f(t,T), \quad T(0) = T_0 ,$$

where $f(t,T) = -k\{T(t) - (T_M + \alpha t)\} + H_c$, then Euler's method, using interval h, would give

$$T_{k+1} = T_k + hf(t_k, T_k) ,$$

with error $O(h^2)$, where $T_k = T(kh)$. A simple computer program could easily be written to find the temperature T within the room after a certain time interval for prescribed values of k, T_M, α and H_c. Clearly such a model would only give reasonable results over a limited time period.

In equation (2.3) the terms $-k(T - T_\alpha)$ and $H(t)$ may fall into either category on the right hand side of equation (2.1), depending on $T > T_\alpha$ or $T < T_\alpha$, and whether $H(t)$ involves heating (e.g., solar or furnace) or cooling (air conditioner). Sometimes the physical quantity involved will enter/leave the 'black-box' R carried along by another substance (e.g., pollutants in a stream entering and leaving a river). At other times it will do so in pure form (e.g., pollutants dumped directly from a steelworks into a lake). In order to formulate simple mathematical models for such problems we assume that the substance is mixed homogeneously throughout the spatial region R. If not, we would have to introduce spatial independent variables in addition to time, and so our models would involve partial differential equations.

Later in this Chapter we will consider both the mathematical formulation and the full solution of other compartment problems. In the next section we develop a mathematical model for a single tank system by considering the inward and outward flow rates. We then extend the model to include the effect of introducing pollutant into the single tank.

2.3 Model for a Single Tank System

We first develop a mathematical model for the flow of water in and out of a single tank as shown in Figure 2.2.

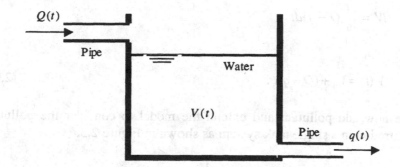

Figure 2.2 Single tank system with pure water.

Let
   $V(t) =$ volume of water in the tank at time t (m^3)
   $Q(t) =$ flow rate into the tank at time t (m^3/sec)
   $q(t) =$ flow rate out of tank at time t (m^3/sec) .

Now consider the change in the situation from time t to time $t + \Delta t$.
   At time t, volume $= V(t)$
   At time $t + \Delta t$, volume $V(t + \Delta t)$
   Volume flowing into the tank in time $\Delta t = Q\Delta t$
   Volume flowing out of the tank in time $\Delta t = q\Delta t$.

New volume at time $t + \Delta t =$ old volume at time $t +$ volume flowing in
                                     $-$ volume flowing out .

$$\therefore \quad V(t+\Delta t) = V(t) + Q\Delta t - q\Delta t$$

$$\therefore \quad \frac{V(t+\Delta t)}{\Delta t} = Q - q$$

In the limit as $\Delta t \to 0$, we have

$$\frac{dV}{dt} = Q - q, \tag{2.6}$$

i.e., rate of change of volume of water $=$ rate at which water flows in
$-$ rate at which water flows out .

In the case where flow rates Q and q are constant and the initial volume $V(0) = V_0$, we have

$$\int_{V_0}^{V} dV = \int_0^t (Q - q) dt$$

$$\therefore \quad V(t) = V_0 + (Q - q)t . \tag{2.7}$$

We now add pollutant and extend the model to consider the pollutant concentration in a single tank system as shown in Figure 2.3.

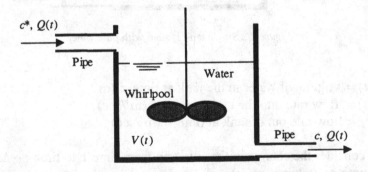

Figure 2.3 Single tank system with pollutant.

Assume that water flows in and out at Q (m³/sec), i.e., the volume V of water in the tank is constant. Let $c(t)$ denote the concentration of pollutant in the tank at time t and assume pollutant enters the tank with concentration c^* (gm/m³).

In order to develop a simple mathematical model it is important to make the main assumption that the solution is "well mixed". In order to avoid having to account for spatial distribution of the pollutant (i.e., different concentrations at differential points in the tank), we assume instantaneous and uniform mixing. This means that the concentration of pollutant in the outflow is also $c(t)$.

To answer the question: "How does the pollutant concentration in the tank vary with time?", we use the conservation of mass law.

Total mass of pollutant in the tank at time $t = c(t)V$

Total mass of pollutant in the tank at time $t + \Delta t = c(t + \Delta t)V$

Mass flowing in $= c^* Q \Delta t$

Mass flowing out $= c(t)Q\Delta t$.

From the mass balance, we have

new mass of pollutant = old mass + mass in − mass out

$$\therefore \qquad c(t + \Delta t)V = c(t)V + c^* Q \Delta t - c(t)Q\Delta t$$

$$\therefore \qquad \frac{c(t + \Delta t) - c(t)}{\Delta t} = \frac{Q}{V}(c^* - c) .$$

In the limit as $\Delta t \to 0$, we have

$$\frac{dc}{dt} = \frac{Q}{V}(c^* - c) . \tag{2.8}$$

Assuming c^*, Q and V are constant and that $c(0) = c_0$, we have

$$\int_{c_0}^{c} \frac{dc}{c^* - c} = \frac{Q}{V} \int_0^t dt$$

$$\therefore \quad \left[-\ln(c^* - c) \right]_{c_0}^{c} = \frac{Q}{V} t$$

$$\therefore \quad \frac{c^* - c}{c^* - c_0} = e^{-Qt/V}$$

$$\therefore \quad c(t) = c^* - (c^* - c_0)e^{-\frac{Qt}{V}} . \tag{2.9}$$

Note that, as a check, we have when $t = 0$, $c(0) = c_0$. Also as $t \to \infty$, $c(t) \to c^*$.

2.4 Well-Posed Problems

We now consider more closely the solution of equation (2.6). Firstly, we must attend to some details to ensure that the problem is well-posed, i.e., that all of the mathematical details are properly stated. Equation (2.6) has three variables that could conceivably be functions of time, namely, $V(t)$, $Q(t)$ and $q(t)$. This means that we have three unknowns and one equation, so that two more mathematical relationships are required. For example, if we assume that the outlet pipe has a valve which regulates the flow, we can use a typical valve equation, which for gravity flow can be written as

$$q = C\sqrt{h} \qquad\qquad (2.10)$$

where C is a constant for a particular valve, and h is the height of liquid in the tank. The problem with equation (2.10) is that if we substitute it into (2.6) we have introduced another unknown, h, so that we still have three unknowns, $V(t)$, $Q(t)$ and $h(t)$. However, it is important to realize that $V = ah$, where A is the tank cross-sectional area (assumed constant), and therefore equation (2.6) gives

$$A\frac{dh}{dt} = Q - q. \qquad\qquad (2.11)$$

It is best not to combine equations (2.10) and (2.11) since, in general, as the mathematical models (sets of equations) for flow systems become more complicated, we find that systems of equations are more convenient to use in computing solutions.

Finally, we assume that the incoming flow rate, Q, is a known function of time (e.g., a given constant). This means that all three variables in equations (2.10) and (2.11) are now defined mathematically, and so we might expect to obtain a solution, in this case $h(t)$. However, there is one other detail we must consider before we have a well-posed problem. We must specify an initial condition for the problem:

$$h(0) = h_0 \qquad\qquad (2.12)$$

where h_0 is a given constant at $t = 0$.

Equations (2.10), (2.11) and (2.12) now constitute the mathematical model for the single tank system of Figure 2.2, and we can proceed to a solution. Here we have purposely developed this model in excessive detail as these steps will apply to all models and are summarized below:

(1) Formulate a set of equations based on a set of assumptions (e.g., constant density) and certain conservation principles (e.g., conservation of mass).

(2) Check to ensure that the units of each term in a given equation are the same as for all other terms in the same equation (e.g., cm^3/sec).

(3) Compute the number of unknowns and, if necessary, add more equations to ensure that the number of equations equals the number of unknowns (e.g., equation (2.10) and $V = Ah$).

(4) Specify an initial condition for each time derivative in a set of first order ordinary differential equations (e.g., $h(0) = h_0$ for dh/dt).

2.5 A Nonisothermal Holding Tank

We now extend the work in Section 2.4 to consider the heating of a liquid in a stirred tank as illustrated in Figure 2.4. The example to be considered illustrates the use of general conservation principles as the starting point for the development of a mathematical model.

In order to calculate the mass of liquid in the tank as a function of time, we apply a total mass balance as done in Section 2.4 to obtain

$$\frac{dV}{dt} = Q - q, \quad V(0) = V_0, \quad h = \frac{V}{A}, \quad q = c\sqrt{h}. \tag{2.13}$$

If we also wish to calculate the temperature of the liquid in the tank as a function of time, we apply an energy balance to the liquid, namely,

$$\begin{bmatrix} \text{Time rate of change} \\ \text{of internal, kinetic} \\ \text{and potential energy} \\ \text{inside the system} \end{bmatrix} = \begin{bmatrix} \text{Flow of internal, kinetic,} \\ \text{and potential energy into} \\ \text{the system by convection} \\ \text{and/or diffusion} \end{bmatrix} - \begin{bmatrix} \text{Flow of internal, kinetic,} \\ \text{and potential energy out} \\ \text{of the system by convection} \\ \text{and/or diffusion} \end{bmatrix}$$

$$+\begin{bmatrix}\text{Rate of heat added to the} \\ \text{system by conduction,} \\ \text{convection, radiation,} \\ \text{and chemical reaction}\end{bmatrix} - \begin{bmatrix}\text{Rate of work done} \\ \text{by the system on} \\ \text{the surroundings}\end{bmatrix} \qquad (2.14)$$

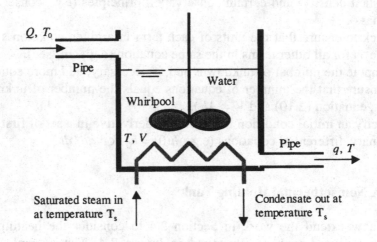

Figure 2.4 Nonisothermal holding tank

The specific mathematical terms are

$$\frac{d}{dt}\{V\rho(u+K+P)\} = Q\rho(u_0+K_0+P_0) - q\rho(u+K+P)$$

$$+UA_T(T_S-T) + \frac{Qp_0}{J} - \frac{qp}{J} \qquad (2.15)$$

where
V = volume of liquid (cm^3)
ρ = liquid density (gm/cm^3)
u = specific internal energy (cal/gm)
K = kinetic energy per unit mass (cal/gm)
P = potential energy per unit mass (cal/gm)
Q , q = liquid flow rates into and out of the tank (cm^3/sec)
U = overall heat transfer coefficient (cal/sec-cm^2- 0° C)
A_T = heat transfer area (cm^2)
T_S = steam temperature (0° C)
T = liquid temperature (0° C)
p_0 , p = external and liquid pressures (dyn/cm^2)

J = conversion factor between mechanical energy and thermal energy (dyn-cm/cal).

The units of each term in equation (2.15) are cal/sec and have physical meaning for each of the terms. For example, the left hand side represents the cal/sec accumulating in the liquid in the tank ($V\rho$ is the mass of liquid). $Q\rho(u_0 + K_0 + P_0)$ and $q\rho(u + K + P)$ are the cal/sec flowing into and out of the tank, respectively. $UA_T(T_s - T)$ is the cal/sec flowing into the liquid as a result of heat transfer, and Qp_0/J and qp/J are the cal/sec of the work being done by the surroundings on the liquid and the work being done by the liquid, respectively.

Equation (2.15) may be simplified by assuming that the kinetic and potential energies are negligible in comparison with the internal energy, i.e., $K \ll u$ and $P \ll u$. This means that

$$\frac{d}{dt}(Vu) = Q\left(u_0 + \frac{p_0}{e^J}\right) - q\left(u + \frac{p_0}{e^J}\right) + \frac{UA_T(T_s - T)}{\rho}. \tag{2.16}$$

Equation (2.16) would give the product Vu as a function of t if integrated numerically. But we are interested in the liquid temperature, T, as a function of t. Hence we must introduce a relationship between u and T.

The internal energy is given by

$$u = C_v(T - T_r), \tag{2.17}$$

where C_v is the specific heat at constant volume and T_r is a reference temperature. Similarly, the enthalpy is given by

$$H = u + \frac{p}{\rho} = C_p(T - T_r), \tag{2.18}$$

where C_p is the specific heat at constant pressure and, for liquids, $C_p \cong C_v$. Therefore equation (2.16) can be written in terms of temperature as

$$\frac{d}{dt}\{V(T - T_r)\} = Q(T_0 - T_r) - q(T - T_r) + \frac{UA_T(T_s - T)}{\rho C_p}. \tag{2.19}$$

Finally, equation (2.19) can be simplified further by taking $T_r = 0$ to give

$$\frac{d}{dt}(VT) = QT_0 - qT + \frac{UA_T(T_S - T)}{\rho C_p}. \tag{2.20}$$

It is important to note that in the process of deriving equation (2.20) we stated a series of assumptions which were used to justify dropping a number of terms in the more general equation (2.15). This explicit statement of the assumptions made in deriving the equations of a mathematical model is quite important to appreciate their limitations and thus avoid using them incorrectly, i.e., using the equations under conditions for which the assumptions are not valid. Also, as each equation of a mathematical model is stated, the units of all of the terms in the equation should be checked to ensure that they are the same throughout the equation.

Examples to illustrate are given below:
(1) If the flow rates Q and q in equation (2.20) have the time units of sec^{-1}, then the heat transfer coefficient, U, should also have the units of sec^{-1}. The checking of time units is a particular requirement of dynamic models.
(2) On multiplying equation (2.20) across by ρC_p, if the derivative on the left hand side, written as $d(V\rho C_p T)/dt$, has the units of cal/sec, each of the terms on the right hand side should also have the units of cal/sec.

Another important point is that the physical meaning of the net units for each term should be understood. For example, cal/sec in the derivative refers to the number of cal/sec accumulating in the tank, cal/sec in each of the two flow terms refers to the number of cal/sec flowing into or out of the tank, and cal/sec in the heat transfer term refers to the number of cal/sec entering or leaving the tank as a result of the temperature difference between the tank liquid temperature and the steam in the heating coil.

The following additional points should be noted concerning equation (2.20):
(1) On integrating $d(VT)/dt$ we obtain $V(t)T(t)$ not $T(t)$. So in order to obtain $T(t)$ we must divide the solution of equation (2.20) by $V(t)$, i.e.

$$T(t) = \frac{V(t)T(t)}{V(t)}$$

where $V(t)$ is obtained from integrating equation (2.13).
(2) The initial condition required by equation (2.20) is $V(0)T(0)$ not $T(0)$. So, in order to start the numerical integration of equation (2.20), we must use $V(0)$, i.e., the initial condition for equation (2.13)   multiplied   by   the

initial temperature, $T(0)$.

2.6 Multi-Compartment Problems

In Sections 2.2 and 2.3 we explored simple linear problems, involving a single compartment (a room/tank) and the time rate of change of some specified quantity (heat energy/pollutant concentration) within that compartment as it interacts with its environment. Also, in Section 2.5, we restricted our attention to a model involving a single holding tank. We now expand on these ideas to allow for more than one compartment.

Consider the simplest generalization, to that of a two-compartment problem. For a two-tank problem, we might represent this diagrammatically as shown in Figure 2.5.

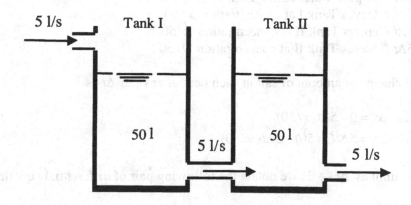

Figure 2.5 System of two tanks.

Test Example 2.1

The two tanks are connected as shown in Figure 2.5 with each tank containing 50 litres of water. Tank I contains 20 grams of dissolved salt, and Tank II contains 4 grams. Pure water flows into the system at 5 litres/sec and a mixture flows out at the same rate.

What is the greatest amount of salt that is ever present in Tank II ?

Solution

Let x = amount of salt in Tank I at time t, y = amount of salt in Tank II at time t. We are given $x(0) = 20$, $y(0) = 4$.

To formulate a mathematical model we make the following assumptions:

(1) Assume liquid is flowing as indicated in Figure 2.5. The concentration of salt in each tank in g/litre is $x/50$ and $y/50$.

(2) Assume mixing is slow enough to allow us to use $x/50$ as concentration of salt in water leaving Tank I while adding to Tank II   but fast enough that we need not worry about how long it takes for   the   salt   to   dissolve uniformly throughout the tank for the next step.
(These assumptions may seem inconsistent).

To obtain the governing differential equations we again consider a short interval of time Δt.

$5\Delta t$ ℓ of pure water enters Tank I
$5\Delta t$ ℓ leaves Tank I at concentration $x/50$
$5\Delta t$ ℓ enters Tank II at concentration $x/50$
$5\Delta t$ ℓ leaves Tank II at concentration $y/50$.

Total change in amount of salt in each tank after time Δt is

I $\Delta x = 0 - 5\Delta t(x/50)$
II $\Delta y = 5\Delta t(x/50) - 5\Delta t(y/50)$.

In the limit as $\Delta t \to 0$, we obtain the following pair of differential equations

$$\frac{dx}{dt} = -0.1x. \tag{2.21}$$

$$\frac{dx}{dt} = 0.1x - 0.1y. \tag{2.22}$$

The objective is to find the maximum y. On differentiating (2.22) and using (2.21) we obtain

$$\frac{d^2y}{dt^2} = 0.1\frac{dx}{dt} - 0.1\frac{dy}{dt}$$

$$= -0.01x - 0.1\frac{dy}{dt}$$

$$= -0.01 \times 10\left(\frac{dy}{dt} + 0.1y\right) - 0.1\frac{dy}{dt} \qquad \text{using (2.22)}$$

$$= -0.1\frac{dy}{dt} - 0.01y - 0.1\frac{dy}{dt} \ .$$

Therefore we obtain the following second order differential equation

$$\frac{d^2y}{dt^2} + 0.2\frac{dy}{dt} + 0.01y = 0 \ .$$

The auxiliary equation is $m^2 + 0.2m + 0.001 = 0$.

$\therefore \ (m - 0.1)^2 = 0$

$\therefore \ $ Roots are $m = 0.1$ (twice)

$\therefore \ $ General solution is $y(t) = (A + Bt)e^{-0.1t}$.

To find the arbitrary constants A and B we impose the initial conditions. From (2.22)

$$x = 10\left(\frac{dy}{dt} + 0.1y\right)$$

$$= 10\left\{(A + Bt)(-0.1e^{-0.1t}) + Be^{-0.1t} + 0.1(A + Bt)e^{-0.1t}\right\}$$

$x(0) = 20 \Rightarrow 20 = 10(-0.1A + B + 0.1A) = 10B$, $\therefore B = 2$; $y(0) = 4 \Rightarrow A = 4$

$\therefore \ y(t) = (4 + 2t)e^{-0.1t}$.

Now to find y_{max} ,

$$\frac{dy}{dt} = (4 + 2t)(-0.1e^{-0.1t}) + 2e^{-0.1t}$$

$$= (1.6 - 0.2t)e^{-0.1t}$$

$$= 0 \ \text{ for a maximum } .$$

Clearly $t = 8$ sec. We should test this is a maximum by finding

$$\frac{d^2 y}{dt^2} = (1.6 - 0.2t)(-0.1e^{-0.1t}) - 0.2e^{-0.1t}$$

$$= (0.02t - 0.36)e^{-0.1t} \ .$$

When $t = 8$,

$$\frac{d^2 y}{dt^2} = -0.2e^{-0.8t} < 0 \ .$$

Clearly $y(t)$ has a maximum at $t = 8$. Also $y_{max} = 20e^{-0.8} = 8.99 \, g$. Hence the greatest amount of salt that is ever present in Tank II is 8.99g and occurs after 8 sec.

As a check the solution using the computer algebra system MAPLE is given in Figure 2.6.

> System := diff(x(t),t) = -0.1*x(t),diff(y(t),t) = 0.1*x(t) - 0.1*y(t);

$$System := \frac{\partial}{\partial t} x(t) = -.1x(t) \, , \ \frac{\partial}{\partial t} y(t) = .1x(t) - .1y(t)$$

> initial := x(0) = 20, y(0) = 4;

$$initial := x(0) = 20 \, , \ y(0) = 4$$

> solution := dsolve({System,initial},{x(t),y(t)});

$$solution := \left\{ x(t) = 20e^{-.10000t} \, , y(t) = 4e^{-.10000t} + 2.0e^{-.100000t}t \right\}$$

> # Find max-y
> dydt := diff(rhs(solution[2]),t);

$$dydt := 1.60000e^{-.10000t} - .20000e^{-.10000t}t$$

> solve(dydt,{t});
    {t = 8}
> # Check d^2y/dt^2 <0
> ddyddt := diff(dydt,t);

$$ddyddt := -0.36000e^{-.10000t} + .20000e^{-.10000t}t$$

> evalf(subs(t=8,ddyddt));

    -0.89866

> # < 0
> ymax := evalf(rhs(subs(t=8,solution[2])));

   $ymax := 8.9866$

Figure 2.6 Calculation of the maximum amount of salt in Tank II for the two tank
problem using the computer algebra system MAPLE.

2.7 Introduction of Organisms

Many industries are faced with the problem of discharging polluted water. In some cases the pollutant may be pumped into local rivers and lakes. This raises many concerns for the safety of the public and other lifeform as the pollutant may contain dangerous chemicals or toxic substances. As a result local councils are empowered to stipulate that pollutant can only be discharged provided an effluent system is installed which ensures that the pollutant concentration entering rivers or lakes is within a specified approved level under all circumstances.

Biological treatment systems which involve the introduction of organisms to eat up the pollutant are generally considered to be the most reliable. The rate of pollutant digestion by the organisms is assumed proportional to the organism concentration with proportionality constant R_1. Also the rate at which organisms multiply is directly proportional to the pollutant concentration with proportionality constant R_2.

In installing an effluent system to 'clean up' the pollutant, the single tank system shown in Figure 2.7 is the most straightforward. It consists of a well-mixed tank containing a volume V of water which has uniform concentrations of pollutant, $c(t)$, and organisms, $B(t)$. Polluted water, with a concentration c^* flows into the tank at the rate Q and clean water flows out at the same rate.

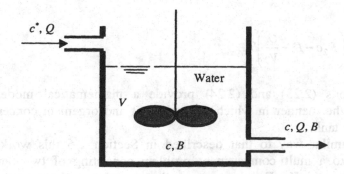

Figure 2.7 Single tank system with pollutant and organisms.

By considering the pollutant level and organism population separately we can devise a mathematical model which involves a pair of ordinary differential equations. As described in Section 2.3, over the time interval $(t, t + \Delta t)$, the mass balance equation for pollutant is

$$\begin{bmatrix} \text{change in pollutant} \\ \text{level in tank} \end{bmatrix} = \begin{bmatrix} \text{pollutant arriving} \\ \text{from inflow} \end{bmatrix} - \begin{bmatrix} \text{pollutant leaving} \\ \text{in outflow} \end{bmatrix} - \begin{bmatrix} \text{pollutant digested} \\ \text{by organisms} \end{bmatrix}$$

Using the notation in Figure 2.7 this leads to

$$V\{c(t + \Delta t) - c(t)\} = c^* Q \Delta t - c(t) Q \Delta t - R_1 c(t) B(t) V \Delta t.$$

In the limit as $\Delta t \to 0$, we obtain

$$\frac{dc}{dt} = (c^* - c)\frac{Q}{V} - R_1 c B. \tag{2.23}$$

The corresponding balance equation for the organisms is

$$\begin{bmatrix} \text{change in organisms} \\ \text{population in tank} \end{bmatrix} = \begin{bmatrix} \text{organisms created} \\ \text{from reproduction} \end{bmatrix} - \begin{bmatrix} \text{organisms lost} \\ \text{via death} \end{bmatrix} - \begin{bmatrix} \text{organisms washed} \\ \text{away in the outflow} \end{bmatrix}$$

which leads to

$$V\{B(t + \Delta t) - B(t)\} = R_2 c(t) B(t) V \Delta t - D B(t) V \Delta t - Q B(t) \Delta t$$

where D is the associated death rate constant. In the limit as $\Delta t \to 0$, we obtain

$$\frac{dB}{dt} = \left(R_2 c - D - \frac{Q}{V} \right) B. \tag{2.24}$$

Equations (2.23) and (2.24) provide a mathematical model which describes the manner in which the pollutant and organism concentrations vary in the tank.

In a similar way to that described in Section 2.6 this work can be extended to a multi-compartment problem consisting of two tanks with volumes V_1 and V_2. For this case, pollutant and organism concentrations would be c_1, B_1 for tank I and c_2, B_2 for tank II. The outflow from tank I would form the inflow to tank II. Also there is no reason to suppose that the growth and digestive behavior of the organisms are any different to that of the single tank system. As before, by carrying out a mass balance, we can describe the behavior of the pollutant and organism concentrations in each of the tanks by, in this case, a system of four ordinary differential equations. Clearly the mathematical model for tank I is equivalent to that for the single tank system.

Clearly, all the work in this chapter could easily be extended to multi-compartment problems consisting of a larger number of compartments than already considered.

2.8 Models Involving Systems of Differential Equations

For many applications in applied mathematics there is not one but several dependent variables, each a function of a single independent variable, usually time. For such compartment problems the modelling process which includes the formulation in mathematical terms frequently leads to a system of differential equations with as many equations as there are dependent variables.

It is important to realise that matrix methods can be used to solve systems of linear differential equations with constant coefficients by the complementary and particular solution method. Any ordinary differential equation or system of ordinary differential equations can be expressed as a system of first order differential equations. To illustrate, we consider

$$\frac{d^n y}{dt^n} = f\left(t, y, \frac{dy}{dt}, \frac{d^2 y}{dt^2}, ..., \frac{d^{n-1} y}{dt^{n-1}}\right).$$

On letting

$$u_1 = \frac{dy}{dt}$$

$$u_2 = \frac{du_1}{dt} = \frac{d^2 y}{dt^2}$$

$$u_3 = \frac{du_2}{dt} = \frac{d^3 y}{dt^3}$$

$$\vdots \qquad \vdots$$

$$u_{n-1} = \frac{du_{n-2}}{dt} = \frac{d^{n-1} y}{dt^{n-1}},$$

we have

$$\frac{dy}{dt} = u_1$$

$$\frac{du_1}{dt} = u_2$$

$$\frac{du_2}{dt} = u_3$$

$$\vdots$$

$$\frac{du_{n-2}}{dt} = u_{n-1}$$

$$\frac{du_{n-1}}{dt} = f(t, y, u_1, u_2, ..., u_{n-1}),$$

which is a system of n first-order equations in the n dependent variables y, u_1, u_2, ..., u_{n-1}. In a similar way we can show that any system of higher order equations may be reduced to a system of first order equations.

We now consider a general system of n first order linear differential equations for the dependent variables x_1, x_2, ..., x_n, namely,

$$\frac{dx_1}{dt} + a_{11} x_1 + a_{12} x_2 + \cdots + a_{1n} x_n = F_1$$

$$\frac{dx_2}{dt} + a_{21} x_1 + a_{22} x_2 + \cdots + a_{2n} x_n = F_2 \qquad (2.25)$$

$$\vdots$$

$$\frac{dx_n}{dt} + a_{n1} x_1 + a_{n2} x_2 + \cdots + a_{nn} x_n = F_n,$$

where the coefficients a_{ij} are given constants and the functions on the right-hand sides $F_1, F_2, ..., F_n$, are given functions of time, t. The system (2.25) can be written in the matrix form

$$\frac{dx}{dt} + Ax = F \qquad (2.26)$$

where $x^T = [x_1, x_2, ..., x_n]$, $F^T = [F_1, F_2, ..., F_n]$ and

$$A = \begin{bmatrix} a_{11} & a_{12} & \cdots & a_{1n} \\ a_{21} & a_{22} & \cdots & a_{2n} \\ \vdots & & & \\ a_{n1} & a_{n2} & \cdots & a_{nn} \end{bmatrix}.$$

The system (2.26) is referred to as a first-order linear matrix differential equation with constant coefficients. To find the complementary solution of (2.26) we make the substitution

$$x = e^{\lambda t} u, \qquad (2.27)$$

where λ is a constant and u is a constant n-dimensional column vector. On substituting (2.27) into (2.26), we obtain

$$\lambda e^{\lambda t} x + A e^{\lambda t} x = 0 \qquad (2.28)$$

or

$$(\lambda I + A) x = 0. \qquad (2.29)$$

It follows that equation (2.29) has a non-trivial solution if and only if

$$\det\left(\lambda\,\mathbf{I}+\mathbf{A}\right)=0\,. \tag{2.30}$$

This is an eigenvalue/eigenvector problem where $-\lambda$ is an eigenvalue of $\mathbf{A}$ and $\mathbf{x}$ is the corresponding eigenvector of $\mathbf{A}$. For problems where $\mathbf{F}\neq\mathbf{0}$ in (2.26) we obtain a particular solution of the matrix equation using the method of variation of parameters. To illustrate the above ideas, we consider two compartment problems one of which is a homogeneous case ($\mathbf{F}=\mathbf{0}$) and the other a non-homogeneous case ($\mathbf{F}\neq\mathbf{0}$). Hence we solve using matrix methods but it is important to note that solutions can be obtained by other methods including D-operator and Laplace transforms.

Test Example 2.2

Some investigations into the distribution, breakdown and synthesis of albumin in animals has led to systems of ordinary differential equations. The compartment model associated with I131 – albumin is shown in Figure 2.8. The process described by this compartment model can be analysed by studying the following system of differential equations:

$$\frac{dx}{dt}=k_2\,y-\left(k_1+k_3\right)x$$

$$\frac{dy}{dt}=k_1\,x-\left(k_2+k_4\right)y$$

$$\frac{dz}{dt}=k_3\,x+k_4y-k_5\,z$$

$$\frac{dw}{dt}=k_5\,z\,.$$

In this biokinetics application, assume that the rate constants have the values $k_1=k_2=k_3=k_4=1$ and $k_5=2$. Also assume that the initial conditions are $x(0)=1$, $y(0)=z(0)=w(0)=0$ and that the terminal conditions are $x=y=z=0$ and $w=1$ as $t\to\infty$. Solve the system.

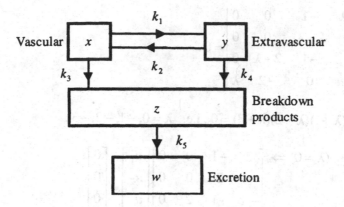

Figure 2.8 Compartment model associated with I131 –albumin.

Solution

On substituting $k_1 = k_2 = k_3 = k_4 = 1$ and $k_5 = 2$, the system becomes

$$\frac{dx}{dt} = -2x + y$$

$$\frac{dy}{dt} = x - 2y$$

$$\frac{dz}{dt} = x + y - 2z$$

$$\frac{dw}{dt} = 2z.$$

We put $\mathbf{x} = e^{\lambda t}\mathbf{u}$ into

$$\frac{d\mathbf{x}}{dt} + \mathbf{A}\mathbf{x} = 0,$$

where

$$\mathbf{A} = \begin{bmatrix} 2 & -1 & 0 & 0 \\ -1 & 2 & 0 & 0 \\ -1 & -1 & 2 & 0 \\ 0 & 0 & -2 & 0 \end{bmatrix}.$$

Then $(\lambda \mathbf{I} + \mathbf{A})\mathbf{u} = 0$ implies that

$$\begin{vmatrix} 2+\lambda & -1 & 0 & 0 \\ -1 & 2+\lambda & 0 & 0 \\ -1 & -1 & 2+\lambda & 0 \\ 0 & 0 & -2 & \lambda \end{vmatrix} = 0.$$

Hence $\lambda(\lambda+1)(\lambda+2)(\lambda+3) = 0$, i.e., $\lambda = 0, -1, -2, -3$.

<u>Case 1</u> $(\lambda = 0) \Rightarrow$ $\begin{bmatrix} 2 & -1 & 0 & 0 \\ -1 & 2 & 0 & 0 \\ -1 & -1 & 2 & 0 \\ 0 & 0 & -2 & 0 \end{bmatrix} \begin{bmatrix} u_1 \\ u_2 \\ u_3 \\ u_4 \end{bmatrix} = \begin{bmatrix} 0 \\ 0 \\ 0 \\ 0 \end{bmatrix}$.

$\therefore$ $\mathbf{u}^T = [0, 0, 0, 1]$ (or any multiple).

<u>Case 2</u> $(\lambda = -1) \Rightarrow$ $\underline{u}^T = [1, 1, 2, -4]$ (or any multiple).

<u>Case 3</u> $(\lambda = -2) \Rightarrow$ $\underline{u}^T = [0, 0, 1, -1]$ (or any multiple).

<u>Case 4</u> $(\lambda = -3) \Rightarrow$ $\underline{u}^T = [1, -1, 0, 0]$ (or any multiple).

Hence

$$\begin{bmatrix} x \\ y \\ z \\ w \end{bmatrix} = C_1 \begin{bmatrix} 0 \\ 0 \\ 0 \\ 1 \end{bmatrix} + C_2 \begin{bmatrix} e^{-t} \\ e^{-t} \\ 2e^{-t} \\ -4e^{-t} \end{bmatrix} + C_3 \begin{bmatrix} 0 \\ 0 \\ e^{-2t} \\ -e^{-2t} \end{bmatrix} + C_4 \begin{bmatrix} e^{-3t} \\ -e^{-3t} \\ 0 \\ 0 \end{bmatrix}.$$

Using the initial conditions, we have

$$C_2 + C_4 = 1$$
$$C_2 - C_4 = 0$$
$$2C_2 + C_3 = 0$$
$$C_1 - 4C_2 - C_3 = 0.$$

The terminal conditions $x = y = z = 0$ as $t \to \infty$ are satisfied without any restriction on the constants C_1, C_2, C_3 and C_4 because of the negative components. The terminal condition $w = 1$ as $t \to \infty$ implies that $C_1 = 1$. Thus $C_1 = 1$, $C_2 = C_4 = 1/2$, $C_3 = -1$. Hence the final solution is

$$x = \frac{1}{2}e^{-t} + \frac{1}{2}e^{-3t}$$

$$y = \frac{1}{2}e^{-t} - \frac{1}{2}e^{-3t}$$

$$z = e^{-t} - e^{-2t}$$

$$w = 1 - 2e^{-t} + e^{-2t}.$$

Next we consider the model of the RLC electrical network shown in Figure 2.9. The mathematical model of this network is obtained from Kirchoff's voltage law which states that "the algebraic sum of all the instantaneous voltage drops around any closed loop is zero, or the voltage impressed on a closed loop is equal to the sum of the voltage drops in the rest of the loop".

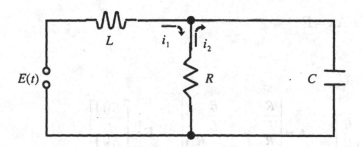

Figure 2.9 RLC electrical network problem.

The left loop yields

$$L\frac{di_1}{dt} + R(i_1 - i_2) = E(t)$$

or

$$\frac{di_1}{dt} + \frac{R}{L}i_1 - \frac{R}{L}i_2 = \frac{E(t)}{L} \tag{2.31}$$

where i_1 is the current in the left loop and i_2 the current in the right loop, $E(t)$ is the electromotive force, and $R(i_1 - i_2)$ is the voltage drop across the resistor R because i_1 and i_2 flow through the resistor R in opposite directions. Similarly for the right loop, we obtain

$$R\,(i_2 - i_1) + \frac{1}{C}\int i_2 \, dt = 0$$

or, by differentiation and division by R,

$$\frac{di_2}{dt} - \frac{di_1}{dt} + \frac{1}{RC}i_2 = 0 \, .$$

On replacing di_1 / dt using (2.31) and ordering, we obtain

$$\frac{di_2}{dt} + \frac{R}{L}i_1 + \left(\frac{1}{RC} - \frac{R}{L} \right)i_2 = \frac{E(t)}{L} \, . \tag{2.32}$$

Equations (2.31) and (2.32) can be written in matrix form

$$\frac{d\mathbf{x}}{dt} + \mathbf{A}\mathbf{x} = \mathbf{F} \, , \tag{2.33}$$

where

$$\mathbf{x} = \begin{bmatrix} i_1 \\ i_2 \end{bmatrix}, \quad \mathbf{A} = \begin{bmatrix} \dfrac{R}{L} & -\dfrac{R}{L} \\ \dfrac{R}{L} & \left(\dfrac{1}{RC} - \dfrac{R}{L} \right) \end{bmatrix}, \quad \mathbf{F} = \begin{bmatrix} \dfrac{E(t)}{L} \\ \dfrac{E(t)}{L} \end{bmatrix} \, .$$

Test Example 2.3

Find expressions for the currents $i_1(t)$ and $i_2(t)$ in the RLC – circuit shown in Figure 2.9 for the case:

$R = 4/5$ ohm, $L = 1$ henry, $C = 1/4$ farad, $E(t) = 4t/5 + 21/25$ volts, $i_1(0) = 1$ ampere, $i_2(0) = -19/5$ amperes.

Solution

For this case the matrix equation (2.33) holds with

$$\mathbf{x} = \begin{bmatrix} i_1 \\ i_2 \end{bmatrix}, \quad \mathbf{A} = \begin{bmatrix} \dfrac{4}{5} & -\dfrac{4}{5} \\ \dfrac{4}{5} & \dfrac{21}{5} \end{bmatrix}, \quad \mathbf{F} = \begin{bmatrix} \dfrac{4t}{5} + \dfrac{21}{25} \\ \dfrac{4t}{5} + \dfrac{21}{25} \end{bmatrix}.$$

First, we find the complementary solution by making the substitution

$$\mathbf{x} = e^{\lambda t}\mathbf{u}$$

into the homogeneous form of equation (2.33), which is

$$\frac{d\mathbf{x}}{dt} + \mathbf{A}\mathbf{x} = \mathbf{0}. \tag{2.34}$$

This yields $(\lambda\,\mathbf{I} + \mathbf{A})\,\mathbf{x} = \mathbf{0}$. For a non-trivial solution

$$\det(\lambda\,\mathbf{I} + \mathbf{A}) = 0 \;,$$

which implies

$$\begin{vmatrix} \lambda + \dfrac{4}{5} & -\dfrac{4}{5} \\ \dfrac{4}{5} & \lambda + \dfrac{21}{5} \end{vmatrix} = 0 \;.$$

Hence

$$\lambda^2 + 5\lambda + 4 = 0$$
$$\therefore \quad (\lambda + 1)(\lambda + 4) = 0$$
$$\therefore \quad \lambda = -1 \quad \text{or} \quad -4.$$

Case 1 $\quad (\lambda_1 = -1) \Rightarrow \quad \begin{bmatrix} -\dfrac{1}{5} & -\dfrac{4}{5} \\ \dfrac{4}{5} & \dfrac{16}{5} \end{bmatrix}\begin{bmatrix} u_1 \\ u_2 \end{bmatrix} = \begin{bmatrix} 0 \\ 0 \end{bmatrix}.$

Therefore $\mathbf{u} = [-4 \; 1]^T$ (or any multiple).

Case 2 $(\lambda_2 = -4) \Rightarrow$
$$\begin{bmatrix} -\dfrac{16}{5} & -\dfrac{4}{5} \\ \dfrac{4}{5} & \dfrac{1}{5} \end{bmatrix} \begin{bmatrix} u_1 \\ u_2 \end{bmatrix} = \begin{bmatrix} 0 \\ 0 \end{bmatrix}.$$

Therefore $\mathbf{u} = [1 \; -4]^T$ (or any multiple).
Hence the complementary solution is

$$\mathbf{x}_c = C_1 \begin{bmatrix} -4e^{-t} \\ e^{-t} \end{bmatrix} + C_2 \begin{bmatrix} e^{-4t} \\ -4e^{-4t} \end{bmatrix}$$
$$= C_1 \mathbf{u}_1 + C_2 \mathbf{u}_2.$$

Next we find a particular solution by replacing the constants C_1 and C_2 by functions $\gamma_1(t), \gamma_2(t)$ so that

$$\mathbf{x}_p = \gamma_1(t)\mathbf{u}_1 + \gamma_2(t)\mathbf{u}_2,$$

where $\mathbf{u}_1$ and $\mathbf{u}_2$ are solutions of the homogeneous equation. On substituting into the system, we obtain

$$\frac{d\mathbf{x}_p}{dt} + \mathbf{A}\mathbf{x}_p = \gamma_1 \frac{d\mathbf{u}_1}{dt} + \gamma_2 \frac{d\mathbf{u}_2}{dt} + \gamma_1' \mathbf{u}_1 + \gamma_2' \mathbf{u}_2 + \mathbf{A}(\gamma_1 \mathbf{u}_1 + \gamma_2 \mathbf{u}_2)$$
$$= \gamma_1 \left(\frac{d\mathbf{u}_1}{dt} + \mathbf{A}\mathbf{u}_1 \right) + \gamma_2 \left(\frac{d\mathbf{u}_2}{dt} + \mathbf{A}\mathbf{u}_2 \right) + \gamma_1' \mathbf{u}_1 + \gamma_2' \mathbf{u}_2$$
$$= \gamma_1' \mathbf{u}_1 + \gamma_2' \mathbf{u}_2.$$

Thus we need to satisfy

$$\gamma_1' \begin{bmatrix} -4e^{-t} \\ e^{-t} \end{bmatrix} + \gamma_2' \begin{bmatrix} e^{-4t} \\ -4e^{-4t} \end{bmatrix} = \begin{bmatrix} \dfrac{4t}{5} + \dfrac{21}{25} \\ \dfrac{4t}{5} + \dfrac{21}{25} \end{bmatrix}$$

or

$$-4\gamma_1' e^{-t} + \gamma_2' e^{-4t} = \frac{4t}{5} + \frac{21}{25}$$

$$\gamma_1' e^{-t} - 4\gamma_2' e^{-4t} = \frac{4t}{5} + \frac{21}{25}.$$

By elimination, we have

$$\gamma_1' = -\frac{1}{15}\left(4t + \frac{21}{5}\right)e^{t}$$

$$\gamma_2' = -\frac{1}{15}\left(4t + \frac{21}{5}\right)e^{4t}.$$

(2.35)

Integrating (3.35) and omitting the constants of integration yields

$$\gamma_1 = -\frac{1}{15}\left(4t + \frac{1}{5}\right)e^{t}$$

$$\gamma_2 = -\frac{1}{15}\left(t + \frac{4}{5}\right)e^{4t},$$

giving the particular solution

$$\mathbf{x}_p = -\frac{1}{15}\left(4t + \frac{1}{5}\right)\begin{bmatrix} -4 \\ 1 \end{bmatrix} - \frac{1}{15}\left(t + \frac{4}{5}\right)\begin{bmatrix} 1 \\ -4 \end{bmatrix}.$$

Hence the general solution is

$$\mathbf{x} = \mathbf{x}_c + \mathbf{x}_p$$

$$= C_1\begin{bmatrix} -4e^{-t} \\ e^{-t} \end{bmatrix} + C_2\begin{bmatrix} e^{-4t} \\ -4e^{-4t} \end{bmatrix}$$

$$-\frac{1}{15}\left(4t + \frac{1}{5}\right)\begin{bmatrix} -4 \\ 1 \end{bmatrix} - \frac{1}{15}\left(t + \frac{4}{5}\right)\begin{bmatrix} 1 \\ -4 \end{bmatrix}.$$

We find C_1 and C_2 using the initial conditions $i_1(0) = 1$, $i_2(0) = -19/5$ which lead to

$$-4C_1 + C_2 = 1$$

$$C_1 - 4C_2 = -4.$$

Hence $C_1 = 0$, $C_2 = 1$ and so the currents $i_1(t)$, $i_2(t)$ are given by

$$i_1(t) = e^{-4t} + t$$

$$i_2(t) = -4e^{-4t} + \frac{1}{5}.$$

2.9 Exercises

1. Formulate a mathematical model to represent the temperature change in a single room over a limited time period. State clearly any assumptions made. Simplify the model under the following assumptions:
(a) Assume the exterior temperature $T_\alpha(t)$ is constant.
(b) Assume the heating (or cooling) rate of the furnace (air-conditioning) is approximately constant while it is operating.
Solve the model to obtain a mathematical expression for the temperature $T(t)$ in the room.

2. A single tank initially contains 60 litres of pure water. Brine containing salt with concentration of 0.1 kilogram/litre enters the tank at the rate of 2 litres/minute. The well-mixed solution leaves the tank at b litres/minute. The tank is empty after exactly 1 hour. Develop a mathematical model and determine b and the maximum concentration of salt.

   State clearly any assumptions made. Check your results using MAPLE and provide a graph showing the concentration of salt versus time.

3. A swimming pool holds 10,000 gallons of water. It can be filled at the rate of 100 gal/min and emptied at the same rate. At the present moment the pool is filled, but there are 20 pounds of impurity dissolved in the water. For safety, this must be reduced to less than 1 pound. It would take 200 minutes to empty the pool completely and refill it, but during part of this time the pool could not be used.

   Formulate a mathematical model, stating all assumptions clearly, and determine how long it would take to restore the pool to safe condition if at all times the pool must be at least half full.

4. Assume the mathematical model described by equations (2.13) and (2.20) for a nonisothermal holding tank as illustrated in Figure 2.4, find values of the temperature $T(t)$ up to $t = 50$ using Euler's method with time step 1. Arrange for the computed results to be printed out at intervals of $t = 5$. You may assume the following values for the parameters:
   $Q = 5,000$, $T_0 = 25$, $U = 100$, $A_T = 2,000$, $T_S = 120$, $\rho = 1$, $C_p = 1$,

$C_v = 1$, $A = 10,000$.

As mentioned in Section 2.5 first solve equation (2.13) to obtain $V(t)$ and then solve equation (2.20) to obtain $V(t)T(t)$. Then obtain $T(t)$ using

$$T(t) = \frac{V(t)T(t)}{V(t)}$$

You may assume the following initial conditions:
$V_0 = V(0) = hA = 50 \times 10,000 = 500,000$
$V_0T_0 = V(0)T(0) = 500,000 \times 25 = 12,500,000$.

5. Suppose that Tanks I and II are placed as shown below and provided with inlet and outlet pipes as indicated.

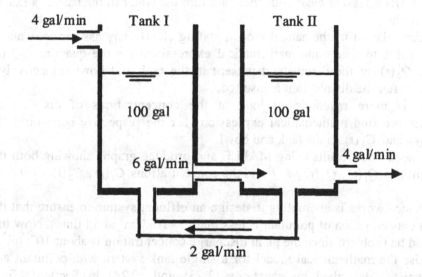

Figure 2.10 System of two tanks with inlet and outlet pipes.

Each tank contains 100 gallons of liquid, and the flow capacity of each pipe is as shown on the diagram. The left-hand pipe admits pure water to Tank I. Suppose that at the moment that all pipes are opened, Tank II contains 10 grams of a toxic substance such as rat poison and Tank I contains only pure water.

Formulate a mathematical model stating clearing any assumptions made. Determine how long it will be until the liquid flowing out of the final outlet pipe on the right is virtually free of the poison, i.e., until the level of concentration decreases to 0.001 g/gal.

Check your results using MAPLE.

6. (a) Use MAPLE in Exercise 5 to produce graphs of the concentrations of toxic substance in Tanks I and II versus time.

(b) Show how these graphs would change if we had started with 10 grams of toxic substance in Tank II and 50 grams in Tank I. Assume everything else stays the same.

(c) Under the circumstances holding in part (b), what would be the effect of changing the flow in the two pipes connecting Tanks I and II from 2 gal/min and 6 gal/min to 12 gal/min and 16 gal/min?

7. The ball valve of the water tank in a toilet is leaking, allowing water to run into the 25 litre tank at a rate of 1 litre/hour. The water flows down the overflow pipe and into the toilet bowl (volume 10 litres) before flowing out into the sewer. A 100 ml bottle of disinfectant is placed in the water tank, and starts releasing cleaner/disinfectant into the water in the tank at a rate of 0.1 ml/hr.

Formulate a mathematical model, stating clearly any assumptions made, and solve to determine mathematical expressions for the quantities $Q_1(t)$ and $Q_2(t)$ of the disinfectant present in the tank and bowl respectively t hours after the disinfectant is inserted.

It is more revealing to look at the concentrations of disinfectant. Therefore find mathematical expressions for the respective concentrations $C_1(t)$ and $C_2(t)$ in the tank and bowl.

Check your results using MAPLE and produce graphs showing both the quantities $Q_1(t)$, $Q_2(t)$ vs. t and the concentrations $C_1(t)$, $C_2(t)$ vs. t.

8. A steelworks is attempting to design an effluent system to ensure that the exit concentration of pollutant is less than $5 \times 10^{-4}/m^3$ at all times. Now this could be problem since the plant discharge concentration is about $10^{-2}/m^3$.

Use the mathematical model for a single tank system with pollutant and organisms described by equations (2.23) and (2.24) in Section 2.7 to determine if a single well-mixed tank with volume 10,000 m^3 would be adequate to meet the above criterion assuming a flow rate through the system of 10 m^3/hour. You may assume the following constants of proportionality:

Birth rate constant, $R_2 = 1.2625$/hour
Death rate constant, $D = 10^{-5}$/hour
Rate constant, $R_1 = 0.1$/hour

By considering the steady state solutions of the differential equations for the pollutant concentration, $c(t)$, and the organism concentration, $B(t)$, determine the volume of a single tank necessary to meet the above criteria.

9. Formulate a mathematical model for a two tank system to deal with the

problem in Exercise 8. As for the single tank system consider the steady state solution for some basic design criteria. Then, by undertaking some numerical experiments, decide whether it would be possible to use a two tank system with smaller combined tank volume than for the single tank system to achieve the same results.

10. Find the currents at any time t in the electrical network shown in the figure below for the case:

$R_1 = 1$ ohm, $R_2 = 4$ ohms, $L = 0.5$ henry, $C = 1$ farad, $E = 100$ volts.

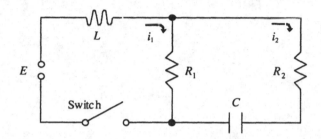

11. Find the currents at any time t in the electrical network shown in the figure below, assuming that they are all zero at time $t = 0$.

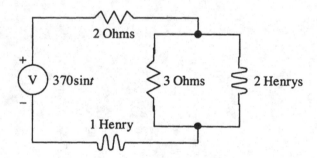

Chapter 3

MODELS IN DYNAMICS AND VIBRATION.

3.1 Elementary Concepts and Results

Let P be a particle of mass m. Denote its position in space as a function of time by the position vector $\mathbf{r}(t)$, measured with respect to some inertial coordinates. Then the velocity of P is defined by

$$\mathbf{v} = \frac{d\mathbf{r}}{dt}$$

and its acceleration is

$$\mathbf{a} = \frac{d^2\mathbf{r}}{dt^2}.$$

The starting point in modelling the motion of the particle is Newton's second law

$$\mathbf{f} = m\mathbf{a}, \tag{3.1}$$

where $\mathbf{f}$ is the resultant of all forces acting on P.

The position of the particle is determined by three parameters. Let $\mathbf{i}$, $\mathbf{j}$ and $\mathbf{k}$ be three unit vectors in the directions of x, y, and z axes of a *Cartesian system*, respectively, as shown in Figure 3.1(a). Then the position, velocity and acceleration of the particle in rectangular coordinates are given by

$$\mathbf{r}(t) = x\mathbf{i} + y\mathbf{j} + z\mathbf{k}, \tag{3.2a}$$

$$\mathbf{v}(t) = \dot{\mathbf{r}}(t) = \dot{x}\mathbf{i} + \dot{y}\mathbf{j} + \dot{z}\mathbf{k}, \tag{3.2b}$$

$$\mathbf{a}(t) = \ddot{\mathbf{r}}(t) = \ddot{x}\mathbf{i} + \ddot{y}\mathbf{j} + \ddot{z}\mathbf{k}, \tag{3.2c}$$

where dots denote derivatives with respect to time.

The *cylindrical coordinates*, shown in Figure 3.1(b), are defined by the three mutually perpendicular unit vectors $\mathbf{e}_r, \mathbf{e}_\theta, \mathbf{k}$. The vector $\mathbf{e}_r$ points from the origin O towards Q, the projection of P on the x-y plane. The length OQ is denoted by r. The second unit vector $\mathbf{e}_\theta$ lies also in the x-y plane. It is perpendicular to $\mathbf{e}_r$ and in the direction of increasing θ. The third unit vector $\mathbf{k}$ is in the z-direction, the same as in the rectangular coordinate system. The position, velocity and acceleration of P, expressed in terms of cylindrical coordinates are

$$\mathbf{r}(t) = r\mathbf{e}_r + z\mathbf{k}, \tag{3.3a}$$

$$\mathbf{v}(t) = \dot{\mathbf{r}}(t) = \dot{r}\mathbf{e}_r + r\dot{\theta}\mathbf{e}_\theta + \dot{z}\mathbf{k}, \tag{3.3b}$$

$$\mathbf{a}(t) = \ddot{\mathbf{r}}(t) = (\ddot{r} - r\dot{\theta}^2)\mathbf{e}_r + (r\ddot{\theta} + 2\dot{r}\dot{\theta})\mathbf{e}_\theta + \ddot{z}(t)\mathbf{k}. \tag{3.3c}$$

The *spherical coordinates* are defined by the three unit vectors $\mathbf{e}_R, \mathbf{e}_\phi, \mathbf{e}_\theta$. The vector $\mathbf{e}_R$ points towards the particle P. The length OP is denoted by R. As in cylindrical coordinates, $\mathbf{e}_\theta$ is a vector in the x-y plane, perpendicular to OQ and in the direction of increasing θ. The vector $\mathbf{e}_\phi$ is in the plane spanned by $\mathbf{e}_R$ and OQ. It is perpendicular to both $\mathbf{e}_R$ and $\mathbf{e}_\theta$, and in the sense of increasing ϕ. In spherical coordinates the position, velocity and acceleration of P are

$$\mathbf{r}(t) = R\mathbf{e}_R, \tag{3.4a}$$

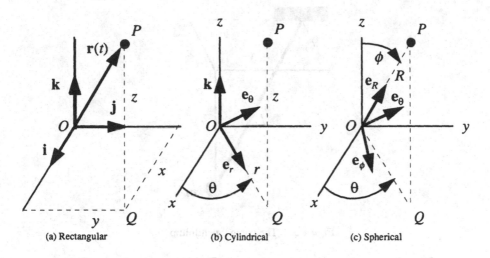

(a) Rectangular (b) Cylindrical (c) Spherical

Figure 3.1 Space coordinate systems

$$v(t) = \dot{\mathbf{r}}(t) = \dot{R}\mathbf{e}_R + R\dot{\phi}\mathbf{e}_\phi + R\dot{\theta}\sin\phi\,\mathbf{e}_\theta, \qquad (3.4b)$$

$$\mathbf{a}(t) = \ddot{\mathbf{r}}(t) = (\ddot{R} - R\dot{\phi}^2 - R\dot{\theta}^2\sin^2\phi)\mathbf{e}_R$$
$$+ (R\ddot{\phi} + 2\dot{R}\dot{\phi} - R\dot{\theta}^2\sin\phi\cos\phi)\mathbf{e}_\phi \qquad (3.4c)$$
$$+ (R\ddot{\theta}\sin\phi + 2\dot{R}\dot{\theta}\sin\phi + 2R\dot{\theta}\dot{\phi}\cos\phi)\mathbf{e}_\theta.$$

Depending on the presence of possible constraints that restrict the motion of P in space, the position of the particle may be determined by less than three parameters. The number of degrees-of-freedom for the particle is determined by the minimal number of independent parameters required to determine its position in space. For example, the spherical pendulum, shown in Figure 3.2, has only two degrees of freedom. Its position is constrained by

$$x^2 + y^2 + z^2 = l^2.$$

The dynamics of a system of q particles $P_1, P_2, ..., P_q$, may be determined by Newton's second law (3.1) applied to each particle, ie.

$$\mathbf{f}_i = m_i \mathbf{a}_i; \quad i = 1, 2, ..., q, \qquad (3.5)$$

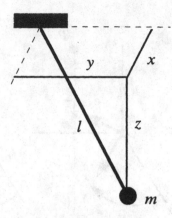

Figure 3.2 The conical pendulum

where $\mathbf{f}_i$ is the resultant force applied on P_i, and m_i and $\mathbf{a}_i$ are its mass and acceleration, respectively. The resultant force $\mathbf{f}_i$ includes the forces applied on P_i by the other q-1 particles. Let $\mathbf{f}_{ij}$ be the force that P_i extracts on P_j. Then Newton's third law states that

$$\mathbf{f}_{ji} = -\mathbf{f}_{ij}, \tag{3.6}$$

and moreover, $\mathbf{f}_{ij}$ and $\mathbf{f}_{ji}$ act along the common line joining the particles. The number of degrees of freedom for the system is defined as the minimal number of independent parameters required to determine the position of all particles in the system. In the presence of s constraints, the number of degrees of freedom for the system is thus $3q$-s.

A rigid body R of density ρ and total mass m may be regarded as a system of particles, constrained to have a fixed distance between each two particles in the body. The vector position of centre of mass G for the rigid body is defined by

$$\mathbf{r}_G = \frac{1}{m} \int_{\mathfrak{R}} \mathbf{r} dm, \ dm = \rho dx \, dy \, dz, \tag{3.7}$$

where $\mathfrak{R}$ denotes the volume of the body (see Figure 3.3(a)). The position of each particle in R can be defined by six parameters, eg. the position of the centre of mass and three Eulerian angles describing the orientation of R. Hence, although R may be regarded as a system of infinite number of particles, with the constraints limiting the motion of the body, the number of degrees of freedom for R is at most six.

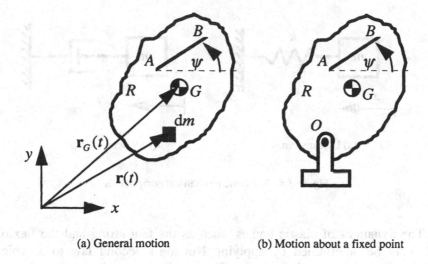

(a) General motion   (b) Motion about a fixed point

Figure 3.3 Motion of a rigid body

We will consider rigid bodies in plane motion only, ie. motion which is restricted to be parallel to a fixed plane. The number of degrees of freedom for a rigid body in plane motion is at most three. Let x-y be the plane of motion for R with some line A-B fixed to the body, as shown in Figure 3.3(a). Denote the location of G by (x_G, y_G), and the angle between A-B and the line $y = 0$ by ψ. Then the three equations of motion that determine the position of each point in R are

$$f_x = m\ddot{x}_G, \quad f_y = m\ddot{y}_G, \quad M_G = I_G\ddot{\psi} \tag{3.8}$$

where f_x and f_y are the x and y components of the resultant force applied on R, M_G is the total moment about G which is applied to R, and

$$I_G = \int_{\Re} \left((x - x_G)^2 + (y - y_G)^2 \right) dm \tag{3.9}$$

is the moment of inertia of R about G. If the body rotates about a fixed point O located at (x_o, y_o), as shown in Figure 3.3(b), then we may use

$$M_O = I_O\ddot{\psi} \tag{3.10}$$

instead of the third equation in (3.8), where M_O and I_O are the applied moment and the moment of inertia of R about O (which can be obtained by replacing the subscript G in (3.9) with O), respectively.

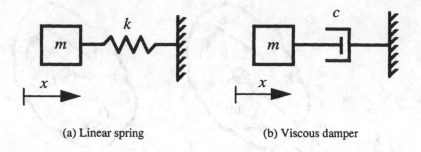

(a) Linear spring   (b) Viscous damper

Figure 3.4 The basic mechanical components

The dynamics of elastic bodies, such as the taut string and the flexible rod, may be determined by applying Newton's second law to a typical infinitesimal element of the body. The resulting equations of motion are then expressed as partial differential equations in time and spatial parameters. Elastic bodies are thus *distributed parameter systems* with an infinite number of degrees of freedom.

Mechanical systems usually consist of masses, springs and dampers. They are the basic mechanical components. The relation between mass and its acceleration is given by Newton's second law (3.1). The spring force is proportional to the product of the spring constant k and the relative displacement of its ends. The spring which is fixed at one end, shown in Figure 3.4(a), applies force F given by

$$F = -kx,\qquad\qquad(3.11)$$

where x is the deflection of the spring from its free length. The force applied by the viscous damper shown symbolically in Figure 3.4(b) is proportional to the product of the damper constant c and the relative velocity of the damper ends. The damper which is fixed at one end, shown in Figure 3.4(b), applies to the mass force given by

$$F = -c\dot{x}.\qquad\qquad(3.12)$$

We now turn our attention to how these principles are used in solving problems in dynamics. In particular, we concentrate on the process of evaluating the model and the form of the equations obtained. We also show how important qualitative properties and characteristic behaviour of the system can be deduced from the derived equations of motion.

3.2 Modelling Single Degree of Freedom Systems

We now consider several dynamic systems and their mathematical models.

A two block system. Consider the two blocks shown in Figure 3.5(a). A constant force F is applied to block B of mass m_B which is connected to block A of mass m_A via an inflexible cable of length l. Assume that friction forces and the mass of the cable are negligible, and that the radius of the pulley is small compared to l, H and D. We wish to determine the equation of the sequential motion for block B before block A hits the pulley.

The system consists of two blocks in plane motion. Hence generally six parameters are needed to determine the position of each point in the blocks. The motion is constrained however by five constraints: the blocks do not rotate, the y-coordinates of blocks are prescribed and the length of the cable connecting the blocks is constant. The system has thus a single degree-of-freedom.

Newton's second law applied to block A in the x-direction gives (see the free body diagram, Figure 3.5(b))

$$T = m_A \ddot{x}_A, \tag{3.13}$$

and for block B we have

$$F - T\cos\alpha = m_B \ddot{x}_B. \tag{3.14}$$

Hence upon eliminating T we obtain

$$m_B \ddot{x}_B + m_A \ddot{x}_A \cos\alpha = F. \tag{3.15}$$

We wish to determine a single differential equation of motion for block B, independent of x_A and its derivatives. The constant length of the cable relates the position of blocks A and B via

$$D - x_A + \sqrt{(x_B - D)^2 + H^2} = l. \tag{3.16}$$

Differentiation with respect to t gives

$$-\dot{x}_A + \frac{x_B - D}{\sqrt{(x_B - D)^2 + H^2}} \dot{x}_B = 0,$$

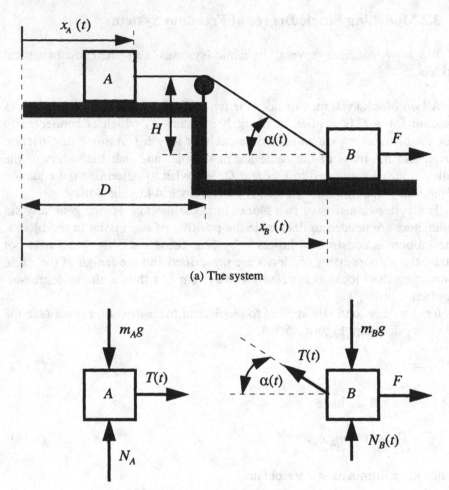

(a) The system

(b) Free body diagrams for the blocks

Figure 3.5 A two-body system

and another differentiation yields

$$\ddot{x}_A = \frac{H^2}{\left((x_B - D)^2 + H^2\right)^{1.5}} \dot{x}_B^2 + \frac{x_B - D}{\sqrt{(x_B - D)^2 + H^2}} \ddot{x}_B .$$

With (3.15) we obtain the differential equation of motion for Block *B*

$$\left(m_B + \frac{m_A(x_B - D)^2}{(x_B - D)^2 + H^2}\right)\ddot{x}_B + \frac{m_A H^2(x_B - D)}{\left((x_B - D)^2 + H^2\right)^2}\dot{x}_B^2 = F. \qquad (3.17)$$

Equation (3.17) together with the initial conditions $x_B(0)$ and $\dot{x}_B(0)$ determine the required motion for block B. The equation of motion in this case is non-linear and it may not be integrable. Hence a closed form solution for $x_B(t)$ may not be found. However, numerical simulation of the solution of (3.17) for specific initial conditions may be easily employed.

We have used the simplified assumptions that there is no friction in the pulley and that the cable is massless. Only with these assumptions does the same tension $T(t)$ appear in both sides of the pulley. In the presence of friction and non negligible mass of the cable the analysis required to determine the motion of the system is more involved. We see that even for a seemingly simple system, with simplified assumptions, the motion may be determined by a complicated non-linear differential equation. It is easily shown, however, that the asymptotic behaviour of the solution of (3.17) when $x_B - D$ is much larger than H is determined by the linear equation

$$(m_A + m_B)\ddot{x}_B = F.$$

The mass-spring-damper system: Consider the mass-spring-damper system moving along the x-axis of a Cartesian coordinate system, shown in Figure 3.6(a). Let x be the displacement of the mass from its static equilibrium position and $f(t)$ be a given external force. Then, ignoring gravity, friction, and with the spring and damper forces given in (3.11)-(3.12) we obtain the free body diagram shown in Figure 3.6(b) for an instant of time where $x > 0$ and $\dot{x} > 0$. Thus Newton's second law applied in the x-direction gives

$$m\ddot{x} + c\dot{x} + kx = f(t). \qquad (3.18)$$

It is easily shown that equation (3.18) still holds for all other possible instants of time, *eg.* when $x > 0$ and $\dot{x} < 0$. The equation of motion for the mass-spring-damper is linear. It is called a *linear time invariant system*, indicating that the coefficients of the differential equation of motion $m > 0$, $c \geq 0$ and $k \geq 0$ are constants, independent of the time parameter. A closed form solution for this system is readily available. The motion and its characteristics for this system will be discussed in the following section.

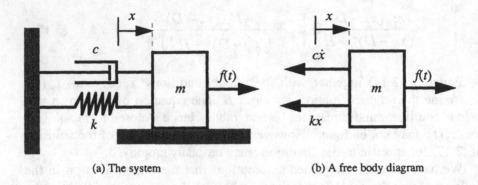

(a) The system   (b) A free body diagram

Figure 3.6 The mass-spring-damper system

A bead on a rotating rod. The bead of mass m slides without friction on the rod, shown in Figure 3.7(a), with a spring of spring constant k and free length L attached to it. The rod rotates in a horizontal plane and forms an angle $\theta(t)$ with the x-axis. We attach a cylindrical coordinate system to the rotating rod, as shown in Figure 3.7(b). The distance of the mass from the fixed point of rotation O is denoted by r. The free body diagram for an instant where $r > L$ is shown in Figure 3.7(b). Two forces are applied on the mass, the spring force $k(r - L)$ in the direction opposite to $\mathbf{e}_r$ and a normal force N which the rod applies on the bead. From Equation 3.3(c) the acceleration of the bead in the $\mathbf{e}_r$-direction is $\ddot{r} - r\dot{\theta}^2$. Thus Newton's second law applied in this direction gives the equation of motion as

$$(\ddot{r} - r\dot{\theta}^2)m = -k(r - L),$$

or equivalently

$$m\ddot{r} + (k - m\dot{\theta}^2)r = kL. \tag{3.19}$$

This equation is linear time variant since the coefficient of r in the differential equation is time dependent when the rotation of the rod is of variable angular speed. If $k < m\dot{\theta}^2$ then this coefficient is negative and the motion of the bead is unstable. The stability of the equation of motion will be discussed in Section 3.4.

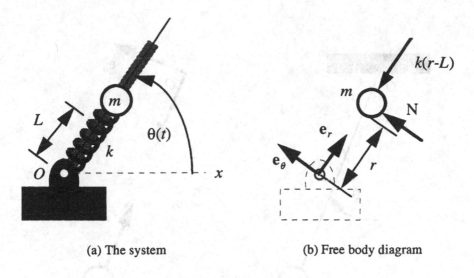

(a) The system   (b) Free body diagram

Figure 3.7 A bead on rotating rod

The simple pendulum. The simple pendulum shown in Figure 3.8(a) vibrates in the *x-y* plane under the influence of gravity. We use cylindrical coordinates with origin at O where $\mathbf{e}_r$ points towards m, as shown in Figure 3.8(b). Then $r = l$ and two forces are applied on the mass, the gravity force mg and the string tension T. The free body diagram for m shows that the component of the gravity force in the $\mathbf{e}_\theta$-direction is $-mg\sin\theta$. For an inextensible string $\dot{r} = \dot{l} = 0$ and it thus follows from equation 3.3(c) that the acceleration in the $\mathbf{e}_\theta$-direction is $l\ddot{\theta}$. Hence Newton's second law applied in this direction gives

$$-mg\sin\theta = ml\ddot{\theta}$$

or

$$l\ddot{\theta} + g\sin\theta = 0 \qquad\qquad (3.20)$$

which is independent of m. This equation describes non-linear oscillations about the equilibrium position $\theta = 0$. For small angle of oscillations we may use the approximation $\theta \cong \sin\theta$ and obtain the linearised equation of motion

$$l\ddot{\theta} + g\theta = 0. \qquad\qquad (3.21)$$

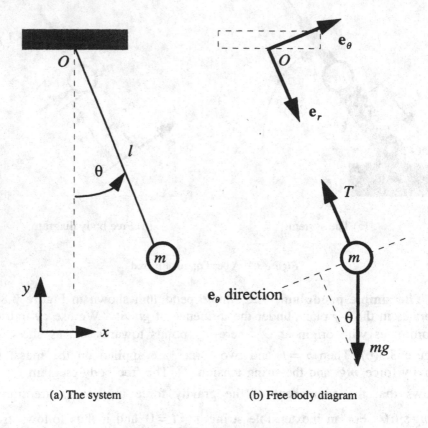

(a) The system   (b) Free body diagram

Figure 3.8 The simple pendulum

A bead on rotating frame. A ring of radius R rotates about the z-axis of a rectangular coordinate system with angular velocity $\dot{\theta}(t)$. A bead of mass m slides on the ring without friction as shown in Figure 3.9(a). We use spherical coordinates with origin at the centre of the ring O, as shown in Figure 3.9(b). The $\mathbf{e}_R$ axis points towards the mass and forms an angle $\phi(t)$ with the z-axis of the Cartesian system. The ring applies a normal force N to the bead in the negative $\mathbf{e}_R$-direction. The free body diagram shows that the component of the gravity force in the $\mathbf{e}_\phi$-direction is $mg\sin\phi$. Since $\dot{R} = 0$ equation 3.4(c) shows that the acceleration of the bead in the $\mathbf{e}_\phi$-direction is $R\ddot{\phi} - R\dot{\theta}^2\sin\phi\cos\phi$ and we have by Newton's second law

$$R\ddot{\phi} - R\dot{\theta}^2\sin\phi\cos\phi - g\sin\phi = 0 . \tag{3.22}$$

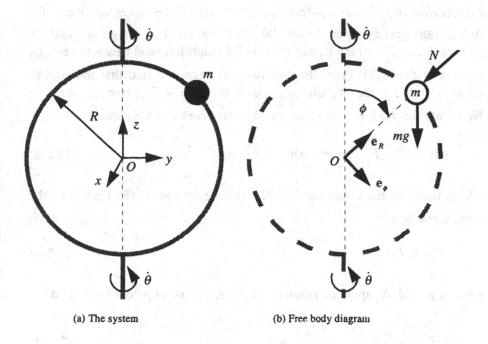

(a) The system   (b) Free body diagram

Figure 3.9 A bead on rotating ring

In the special case where $\dot{\theta}$ is constant there are generally four possible equilibrium positions. The stability of the motion about these points will be discussed in Section 3.4.

The rolling disk. To demonstrate the use of modelling the dynamics of a rigid body in plane motion we consider the disk of radius a, shown in Figure 3.10(a), that rolls without slipping on the circular arc of radius b. Let Q be the point on the disk that is in contact with the ground. Then the no-slip condition implies that the velocity Q vanishes. So far we have dealt with geometrical constraints restricting the position of the system involved. The no-slip condition is a kinematical constraint involving a restriction in terms of velocity. So the rolling disk is indeed a single degree of freedom system. The plane motion of a rigid body which is generally described by three parameters has here two constraints. One geometrical constraint is $x_G^2 + y_G^2 = (b-a)^2$ where x_G and y_G are the coordinates of the centre of the disk with respect to a Cartesian system with origin at O. The other kinematical constraint is $\mathbf{v}_Q = \mathbf{0}$.

We now use cylindrical coordinates with origin O, as shown in Figure 10(b). Then using equation (3.3c) with constant $r = b - a$ we find that the acceleration of G in the $\mathbf{e}_\theta$-direction is $(b-a)\ddot\theta$. The forces applied to the disk are the gravity force, the normal force N applied by the ground and the friction force F_f which causes the no-slip condition (and hence the rolling motion of the disk). From the free body diagram we find that the sum of these forces in the $\mathbf{e}_\theta$-direction is $-mg\sin\theta - F_f$. Hence, applying Newton's second law with respect to the mass centre G we obtain

$$-mg\sin\theta - F_f = m(b-a)\ddot\theta \tag{3.23}$$

which involves the unknown F_f. The third equation of (3.8) gives us the second relation

$$-F_f a = I_G \ddot\psi \tag{3.24}$$

since mg and N apply no moment about G. The no-slip condition yields

$$\ddot\psi = -\frac{b-a}{a}\ddot\theta . \tag{3.25}$$

Substituting (3.24) and (3.25) in (3.23), the equation of motion for the centre of mass of the disk is obtained as

$$(b-a)\left(m + \frac{I_G}{a^2}\right)\ddot\theta + mg\sin\theta = 0. \tag{3.26}$$

If the disk is uniform, ie. its density per unit area is $\rho = m/\pi a^2$, then (3.9) is easily integrable in polar coordinates. By using $dm = \rho r dr d\theta$ and $r^2 = (x - x_G)^2 + (y - y_G)^2$ we obtain

$$I_G = \int\limits_0^{2\pi}\int\limits_0^a \rho r^3 dr d\theta = 2\pi\rho\frac{a^4}{4} = \frac{m}{2}a^2 . \tag{3.27}$$

In this case the equation of motion (3.26) gives

$$\frac{3(b-a)}{2}\ddot\theta + g\sin\theta = 0. \tag{3.28}$$

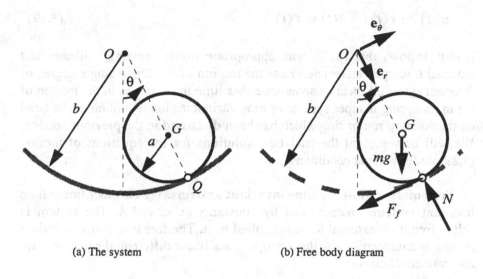

(a) The system   (b) Free body diagram

Figure 3.10 The rolling disk

Comparing with the equation of motion for the simple pendulum (3.21) we find that the motion of G is equivalent to the motion of a pendulum of length $l = 1.5(b-a)$.

In the absence of friction the disk slides on the ground and $F_f = 0$. We obtain immediately from (3.23) that in this case the motion of G is similar to that of a pendulum of length $l = b - a$.

3.3 The Response of Single Degree of Freedom Systems

The motion of a single degree of freedom system is governed by a second order differential equation. The equation of motion is either linear or non-linear. Non-linear differential equations may be integrated numerically by, for example, Runge-Kutta methods, and the solution for specific initial conditions may be simulated. Simulations, however, are not always illuminating. Qualitative properties and characteristics which are usually more important than time history records of a solution to particular initial conditions may not be exposed. Some non-linear differential equations, such as that associated with the motion of the pendulum, may be linearised using certain assumptions. The motion may then be approximated for the interval of time for which the linearisation assumptions are valid, by using the analytical solution of the linear differential equation obtained.

The most general linear second order differential equation has the form

$$m(t)\ddot{x} + c(t)\dot{x} + k(t) = f(t).$$
(3.29)

It thus follows that (3.29) with appropriate mass, damping, stiffness and external force parameters describes the motion of any linear single degree of freedom system. Linear systems are either time invariant, as in the motion of the mass-spring-damper system, or time variant, as in the motion of the bead on the rotating rod or ring which has been discussed in the previous section. We will now present the analytical solutions for the equations of motion (3.29) under various conditions.

Free motion of linear time invariant systems. We consider linear time invariant systems characterised by constants m, c, and k. The system is called free if no external force is applied to it. The free linear time invariant system is characterised by the homogeneous linear differential equation with constant coefficients

$$m\ddot{x} + c\dot{x} + k = 0.$$
(3.30)

When $m > 0$, and $c, k \geq 0$, we may express (3.30) alternatively in the standard form

$$\ddot{x} + 2\zeta\omega_n\dot{x} + \omega_n^2 x = 0$$
(3.31)

where

$$\omega_n = \sqrt{\frac{k}{m}}$$
(3.32)

is called (for a reason that will become apparent soon) the *natural frequency* of the undamped system, and

$$\zeta = \frac{c}{2\sqrt{km}}$$
(3.33)

is called the *damping ratio*. Denote the *damped natural frequency* ω_d by

$$\omega_d = \omega_n\sqrt{1 - \zeta^2}.$$
(3.34)

Then the general solution of (3.31) depends on the value of ζ, and is given by

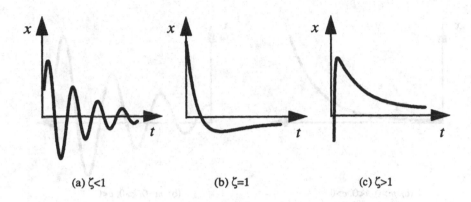

(a) $\zeta < 1$ (b) $\zeta = 1$ (c) $\zeta > 1$

Figure 3.11 The response of free time invariant system

$$x(t) = \begin{cases} A_1 e^{-\zeta \omega_n t} \sin \omega_d t + B_1 e^{-\zeta \omega_n t} \cos \omega_d t; & \zeta < 1 \\ A_2 e^{-\omega_n t} + B_2 t e^{-\omega_n t}; & \zeta = 1 \\ A_3 e^{(-\zeta + \sqrt{\zeta^2 - 1})\omega_n t} + B_3 e^{(-\zeta - \sqrt{\zeta^2 - 1})\omega_n t}; & \zeta > 1 \end{cases} \qquad (3.35)$$

where A_i and B_i, $i = 1, 2, 3$, are arbitrary constants which are determined by the initial conditions $x(0)$ and $\dot{x}(0)$. Denoting the *critical damping* by

$$c_c = 2\sqrt{mk}, \qquad (3.36)$$

then (3.33) can be written as

$$\zeta = \frac{c}{c_c}, \qquad (3.37)$$

and for this reason ζ is called the damping ratio.

It follows from (3.35) that if $\zeta < 1$ then the mass-spring-damper oscillates with damped natural frequency ω_d. We note that $x \rightarrow 0$ as $t \rightarrow \infty$ similar to the motion of physical flexible objects free from external forces. When $\zeta \geq 1$ the motion is non-oscillatory. The motion of the system in these cases for specific initial conditions is illustrated in Figure 3.11. Hence, if the damping coefficient c is less than the critical damping c_c the system moves with damped oscillatory motion, If $c \geq c_c$ there will be no oscillatory motion and the system will approach the steady state exponentially.

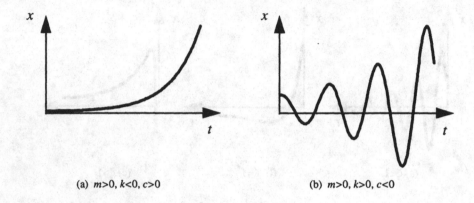

(a) $m>0, k<0, c>0$ (b) $m>0, k>0, c<0$

Figure 3.12 Unstable motion

Consider now the case where $m > 0$ and $k < 0$, as may occur when the bead on the rotating rod (see equation (3.19)) is subject to a large angular velocity. We substitute

$$x(t) = e^{st} \tag{3.38}$$

in (3.30) to obtain the characteristic equation

$$ms^2 + cs + k = 0. \tag{3.39}$$

Hence the *poles* s_1, s_2 of the system, given by

$$s_{1,2} = \frac{-c \pm \sqrt{c^2 - 4km}}{2m}, \tag{3.40}$$

are real, and the general solution of (3.30) is

$$x(t) = Ae^{s_1 t} + Be^{s_2 t}, \tag{3.41}$$

where A and B are arbitrary constants. Regardless of whether c is positive or negative, one of the poles is positive. The response of the system to an arbitrary initial condition therefore increases without bound as $t \rightarrow \infty$ and the system is said to be *unstable*. Such motion is illustrated schematically in Figure 3.12(a).

Using similar analysis we find that when m, $k > 0$ and $c < 0$ the real part of the poles s_1 and s_2 is positive. The motion in this case is unstable. It may be of oscillatory nature as shown in Figure 3.12(b), or non-oscillatory, depending on whether the poles are complex or real.

The impulse response. The impulse of a force $f(t)$ between two instants of time, t_1 and t_2, is defined by

$$\int_{t_1}^{t_2} f(t)dt .$$

(3.42)

The *delta function* $\delta(t - \tau)$, satisfying its definition

$$\begin{cases} \delta(t - \tau) = 0, & t \neq \tau \\ \int_{-\infty}^{\infty} \delta(t - \tau)dt = 1 \end{cases} ,$$

(3.43)

represents a unit impulse. We may think of this force as the limit value of the Boxcar function of unit area shown in Figure 3.13, with $\varepsilon \rightarrow 0$. In physical reality it may represent a certain large force applied during a short period of time about $t = \tau$, such as the force occurring during an impact of colliding bodies.

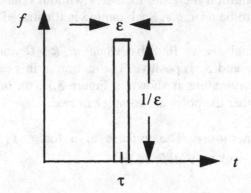

Figure 3.13 Approximate impulse

We now wish to determine the response of the linear time invariant system which is initially at rest to the force $f = \delta(t - \tau)$. We will denote the solution to this problem by $h(t - \tau)$. Then $h(t - \tau)$ is the response of the system to a unit impulse, or in short the *impulse response*. The impulse response is determined by the solution of the differential equation

$$m\ddot{h} + c\dot{h} + kh = \delta(t - \tau),\tag{3.44}$$

with the conditions

$$h = 0, \quad t < \tau,\tag{3.45}$$

$$\dot{h} = 0, \quad t < \tau.\tag{3.46}$$

Obviously (3.44)-(3.46) imply that the system has no motion for $t < \tau$. Let us now determine the dynamics of the system for $t > \tau$. Since no force is applied after the impact, the dynamics of the system after the impact is governed by

$$m\ddot{h} + c\dot{h} + kh = 0, \quad t > \tau,\tag{3.47}$$

$$h(\tau^+) = A,$$ (3.48)

$$\dot{h}(\tau^+) = B$$ (3.49)

where τ^+ represents the time immediately after the application of the impulsive force, and where A and B are some constants representing the displacement and velocity of the mass at the time $t = \tau^+$, respectively. Similarly the instant of time immediately before the impact is denoted by $t = \tau^-$.

We now show how to evaluate A and B. It follows from (3.43) that

$$\int_{-\infty}^{\infty} \delta(t - \tau)dt = \int_{\tau^-}^{\tau^+} \delta(t - \tau)dt = 1.$$ (3.50)

Hence integrating (3.44) with respect to time form $t = \tau^-$ to $t = \tau^+$ gives

$$\int_{\tau^-}^{\tau^+} m\ddot{h}dt + \int_{\tau^-}^{\tau^+} c\dot{h}dt + \int_{\tau^-}^{\tau^+} khdt = \int_{\tau^-}^{\tau^+} \delta(t - \tau)dt = 1$$ (3.51)

Since the position and velocity of the system during the application of the impulse are finite we conclude that

$$\int_{\tau^-}^{\tau^+} m\ddot{h}dt = 1.$$ (3.52)

Equation (3.52) yields

$$m\dot{h}\Big|_{t=\tau^-}^{\tau^+} = m\dot{h}(\tau^+) - m\dot{h}(\tau^-) = m\dot{h}(\tau^+) = 1,$$ (3.53)

since $\dot{h}(\tau^-) = 0$. It thus follows from (3.49) and (3.53) that

$$B = \frac{1}{m}.$$ (3.54)

This important result shows that the velocity gained by the impulse is m^{-1}, which is independent of the spring constant k and the damper constant c. In fact, this is no more than the fundamental principle of impulse and momentum: "*The change in momentum equals the impulse of the acting force*".

To determine the constant A we integrate (3.53) to give

$$\int_{\tau^-}^{\tau^+} m\dot{h}\, dt = \int_{\tau^-}^{\tau^+} dt \qquad\qquad (3.55)$$

and obtain

$$m\dot{h}\Big|_{t=\tau^-}^{\tau^+} = t\Big|_{t=\tau^-}^{\tau^+} = \tau^+ - \tau^- \to 0. \qquad\qquad (3.56)$$

By virtue of (3.45) $h(\tau^-) = 0$ and it thus follows from (3,48) that $A = 0$. In other words, *the mass cannot gain a finite displacement during an infinitesimal period of time*.

The impulse response is thus governed by the initial value problem

$$m\ddot{h} + c\dot{h} + kh = 0, \quad t - \tau > 0$$
$$h(\tau^+) = 0 \qquad\qquad (3.57)$$
$$\dot{h}(\tau^+) = \frac{1}{m}$$

With the time shift $t - \tau \to t$ we have

$$m\ddot{h} + c\dot{h} + kh = 0, \quad t > 0$$
$$h(0) = 0 \qquad\qquad (3.58)$$
$$\dot{h}(0) = \frac{1}{m},$$

which, for $\zeta < 1$, has the solution (see (3.35))

$$h(t - \tau) = \begin{cases} 0, & t < \tau \\ \dfrac{e^{-\zeta\omega_n(t-\tau)}}{m\omega_d}\sin(\omega_d(t - \tau)), & t > \tau \end{cases} \qquad\qquad (3.59)$$

Similar expressions can be obtained via (3.35) for the other cases where $\zeta \geq 1$.

The response to excitation by convolution. Consider an arbitrary function $f(t)$. Then for a small time period $\Delta \tau$ about $t = \tau$ the impulse of $f(t)$ is approximately $f(\tau)\Delta\tau\delta(t-\tau)$, as shown in Figure 3.15. The total impulse of $f(t)$ may thus be expressed by the convoluted sum $\sum f(\tau)\Delta\tau\delta(t-\tau)$. Let $f(\tau)\Delta\tau h(t-\tau)$ be the response of the system due to the impulse $f(\tau)\Delta\tau\delta(t-\tau)$. Then the total response to the excitation $f(t)$ is

$$x(t) = \sum f(\tau)\Delta\tau h(t-\tau).$$

At the limit, when $\Delta\tau \to 0$, this summation is replaced by integration

$$x(t) = \int_0^t f(\tau)h(t-\tau)d\tau. \tag{3.60}$$

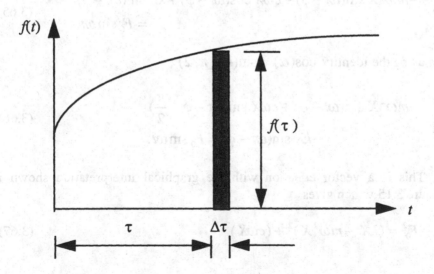

Figure 3.14 Impulse of a force

Since t is a dummy variable in the convolution integral (3.60), we have an alternative expression

$$x(t) = \int_0^t f(t - \tau)h(\tau)d\tau .$$

$$(3.61)$$

The frequency response. If a system is excited by the harmonic force

$$f(t) = F_o \sin \omega t ,$$

$$(3.62)$$

where F_o is constant, then the system is said to be a harmonically excited system. If such excitation is applied to the mass-spring-damper system which is initially at rest then the response is determined by the solution of

$$m\ddot{x} + c\dot{x} + kx = F_o \sin \omega t .$$

$$(3.63)$$

We wish to determine a particular solution of the form

$$x_P(t) = X \sin(\omega t - \phi) ,$$

$$(3.64)$$

where X and ϕ are constants and ω is the frequency of the exciting force. Substituting (3.64) into (3.63) gives

$$-m\omega^2 X \sin(\omega t - \phi) + c\omega X \cos(\omega t - \phi) + kX \sin(\omega t - \phi)$$
$$= F_0 \sin \omega t,$$

$$(3.65)$$

or, using the identity $\cos(\alpha) = \sin(\alpha + \pi/2)$

$$-m\omega^2 X \sin(\omega t - \phi) + c\omega X \sin(\omega t - \phi + \frac{\pi}{2})$$
$$+kX \sin(\omega t - \phi) = F_0 \sin \omega t.$$

$$(3.66)$$

This is a vector equation with the graphical interpretation shown in Figure 3.15 which gives

$$F_0^2 = (kX - m\omega^2 X)^2 + (c\omega X)^2 ,$$

$$(3.67)$$

or

$$X = \frac{F_0}{\sqrt{(k - m\omega^2)^2 + (c\omega)^2}} .$$

$$(3.68)$$

We also have

$$\tan\phi = \frac{c\omega X}{kX - m\omega^2 X} = \frac{c\omega}{k - m\omega^2},$$ (3.69)

and hence

$$\phi = \tan^{-1}\frac{c\omega}{k - m\omega^2}.$$ (3.70)

Alternatively, by using (3.32) and (3.33), we may write X and ϕ in terms of ω_n and ζ, as

$$X = \frac{F_0/k}{\sqrt{\left(1 - \left(\frac{\omega}{\omega_n}\right)^2\right)^2 + \left(2\zeta\frac{\omega}{\omega_n}\right)^2}}$$ (3.71)

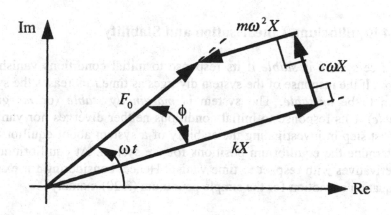

Figure 3.15 Vector plot of Equation (3.66)

and

$$\phi = \tan^{-1} \frac{2\zeta \dfrac{\omega}{\omega_n}}{1 - \left(\dfrac{\omega}{\omega_n}\right)^2}.$$

(3.72)

The general solution of (3.63) is

$$x(t) = x_H(t) + x_P(t)$$

(3.73)

where $x_H(t)$ is the general solution of the associated homogeneous differential equation given by (3.35). Since $x_H \to 0$ for large t the particular solution x_P, given by (3.64) together with (3.71) and (3.72), is the steady state response of the mass-spring-damper system.

It follows from (3.71) that if $c=0$ and the exciting frequency $\omega = \omega_n$ then the amplitude $X \to \infty$. This is the *resonance* phenomenon which yields large amplitude oscillations and mechanical failure in vibrating structures.

3.4 Equilibrium, Linearisation and Stability

A free system is *stable* if its response to initial conditions vanishes as $t \to \infty$. If the response of the system diverges as time t increases the system is said to be *unstable*. The system is *marginally stable* (or *marginally unstable*) if its response to initial conditions neither diverges nor vanishes. The first step in investigating the stability of a system about equilibrium is to determine the equilibrium positions for the system. At equilibrium state the derivatives with respect to time vanish. Hence, considering for example the equation of motion for the simple pendulum (3.20), namely

$$l\ddot{\theta} + g\sin\theta = 0,$$

(3.74)

we find that for the equilibrium state the following equation holds

$$g\sin\theta = 0.$$

(3.75)

Solving (3.75) for θ we obtain two equilibrium points, namely $\theta_1 = 0$ and $\theta_2 = \pi$, associated with the positions of the pendulum shown in Figures

3.16(a) and 3.16(b) respectively. In the second step, after the equilibrium positions have been found, an assumption of small motion about each equilibrium point is made, and a linearised system of equations about the equilibrium position is determined. For the simple pendulum, small motion assumption about $\theta_1 = 0$ leads to the approximation $\theta \cong \sin\theta$. The linear equation of motion associated with (3.74) about $\theta = 0$ is thus

$$l\ddot{\theta} + g\theta = 0. \tag{3.76}$$

With the assumption of small motion $\phi(t)$ about the second equilibrium position, $\theta_2 = \pi$, we substitute $\theta = \pi + \phi$ in (3.74) and obtain

$$l\ddot{\phi} + g\sin(\pi + \phi) = l\ddot{\phi} - g\sin\phi = 0.$$

Then with the approximation $\phi \cong \sin\phi$ we obtain the linear version of (3.74) associated with the motion of the inverted pendulum

$$l\ddot{\phi} - g\phi = 0. \tag{3.77}$$

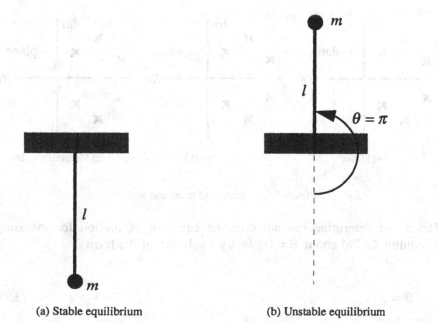

(a) Stable equilibrium   (b) Unstable equilibrium

Figure 3.16 Equilibrium positions for the pendulum

In the third step the stability of the motion is determined by considering the poles of the resulting linear equations of motion. The system is stable if all the poles s_i associated with the differential equation of motion satisfy $\text{Re}(s_i) < 0$. The system is unstable if for at least one pole s_p we have $\text{Re}(s_p) > 0$. If all poles satisfy $\text{Re}(s_i) \leq 0$, where the equality holds for some simple poles, then the motion is marginally stable about the equilibrium position. It is customary in control to describe the stability in a pictorial manner. The complex plane, called also the s-plane, shown in Figure 3.17, is divided by the imaginary axis to right-hand-side and left-hand-side. The equilibrium position is stable if all the poles associated with the equation of motion are strictly in the left-hand-side of the s-plane, as in Figure 3.17(a). If one, or more, pole is located in the right-hand-side of the s-plane the system is unstable, as in Figure 3.17(b). The motion is marginally stable if there are no poles in the right-hand-side of the s-plane, but at least one of the simple poles is located on the imaginary axis, as in 3.17(c).

Note that the poles of a differential equation with real coefficients form a self-conjugate set in the sense that if s is a complex pole of the system then its complex conjugate $\bar{s}$ is another pole of the system.

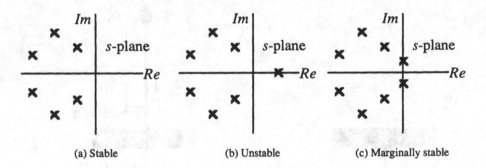

Figure 3.17 Location of poles and stability

Hence, to determine the poles of the equation of motion for the simple pendulum (3.76) about $\theta = 0$, we try a solution of the form

$$\theta = e^{st}. \tag{3.78}$$

Substituting (3.78) in (3.76) yields the characteristic equation

$$s^2 l + g = 0,$$

(3.79)

with poles

$$s_{1,2} = \pm i \sqrt{g/l} \, .$$

These poles are located on the imaginary axis of the s-plane, as shown in Figure 3.18(a). The motion of the pendulum is thus marginally stable about this equilibrium position. The assumption of small oscillation about $\theta = 0$ is thus justified, provided that the initial conditions are small.

Substituting $\phi = e^{st}$ in the equation of motion for the inverted pendulum (3.77) gives

$$s^2 l - g = 0.$$

(3.80)

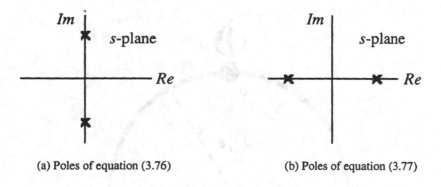

(a) Poles of equation (3.76)   (b) Poles of equation (3.77)

Figure 3.18 Poles of the simple pendulum

The poles in this case are

$$s_{1,2} = \pm \sqrt{g/l} \, ,$$

and the motion of the inverted pendulum is unstable since s_1 is in the right hand side of the complex plane, as illustrated in Figure 3.18(b).

The above results fully confirm our intuition about the motion of the pendulum. The method just described is general and can be applied to more involved cases, as demonstrated by the following example.

Stability of the bead on the rotating frame. We analyse the stability of the bead of the rotating frame system studied in Section 3.2. Suppose $R = 1$ and $\dot{\theta}^2 = \sqrt{2}g$. Then the differential equation of motion for the bead (3.22) is given by

$$\ddot{\phi} - \sqrt{2}g \sin\phi \cos\phi - g \sin\phi = 0 \qquad (3.81)$$

Hence, the equilibrium positions are the angles ϕ_i which are the roots of

$$-\sqrt{2}g \sin\phi \cos\phi - g \sin\phi = 0,$$

or

$$(\sqrt{2}\cos\phi + 1)\sin\phi = 0. \qquad (3.82)$$

There are thus four equilibrium positions

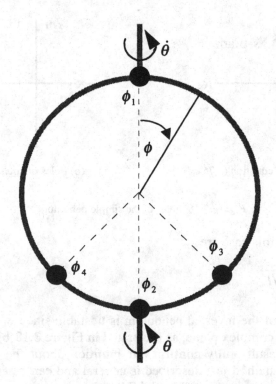

Figure 3.19 Equilibrium positions for the bead

$$\phi_1 = 0, \ \phi_2 = \pi, \ \phi_3 = \frac{3}{4}\pi \ \text{and} \ \phi_4 = -\frac{3}{4}\pi$$

which are shown in Figure 3.19.

For small motion about $\phi_1 = 0$ we use the approximation $\sin\phi \cong \phi$, $\cos\phi \cong 1$ in (3.81) and obtain

$$\ddot{\phi} - (\sqrt{2} + 1)g\phi = 0. \tag{3.83}$$

Since the poles of (3.83) are $s_{1,2} = \pm\sqrt{(\sqrt{2} + 1)g}$, the equilibrium position about $\phi_1 = 0$ is unstable.

For small motion about $\phi_2 = \pi$ we substitute in (3.81) $\phi = \pi + \varphi$, where φ is a small angle, and obtain

$$\ddot{\phi} - \sqrt{2}g\sin(\pi + \varphi)\cos(\pi + \varphi) - g\sin(\pi + \varphi) = 0.$$

Then with the approximation $\sin(\pi + \varphi) \cong -\varphi$, $\cos(\pi + \varphi) \cong -1$ we have

$$\ddot{\phi} + (1 - \sqrt{2})g\phi = 0. \tag{3.84}$$

This equilibrium position, with poles $s_{1,2} = \pm\sqrt{(\sqrt{2} - 1)g}$, is unstable as well.

For small motion about $\phi_3 = 3\pi/4$ we substitute $\phi = 3\pi/4 + \varphi$, where φ is a small angle, in (3.81) and obtain

$$\ddot{\phi} - \sqrt{2}g\sin(\frac{3}{4}\pi + \varphi)\cos(\frac{3}{4}\pi + \varphi) - g\sin(\frac{3}{4}\pi + \varphi) = 0. \tag{3.85}$$

Then with the trigonometric identities

$$\sin(\alpha + \beta) = \cos\alpha\sin\beta + \sin\alpha\cos\beta$$

$$\cos(\alpha + \beta) = \cos\alpha\cos\beta - \sin\alpha\sin\beta$$

we approximate

$$\sin(\frac{3}{4}\pi + \varphi) \cong -\frac{\sqrt{2}}{2}(\varphi - 1), \quad \cos(\frac{3}{4}\pi + \varphi) \cong -\frac{\sqrt{2}}{2}(\varphi + 1)$$

and obtain from (3.85)

$$\ddot{\phi} - \frac{\sqrt{2}}{2}g(\varphi^2 - 1) + \frac{\sqrt{2}}{2}g(\varphi - 1) = \ddot{\phi} + \frac{\sqrt{2}}{2}g(\varphi - \varphi^2) = 0.$$

By omitting the second order term we find the linear version of (3.85)

$$\ddot{\phi} + \frac{\sqrt{2}}{2}g\varphi = 0, \tag{3.86}$$

which describes marginally stable equilibrium oscillations about $\phi_3 = 3\pi/4$. Similar analysis shows that $\phi_4 = -3\pi/4$ is another position of marginally stable motion. With appropriate initial conditions, the bead will thus vibrate about either ϕ_3 or ϕ_4.

3.5 Modelling Systems with Many Degrees of Freedom.

The motion of a single degree-of-freedom system is governed by a second order differential equation. The motion of a system with n degrees of freedom is governed by n second order differential equations. Newton's second law can be used with respect to each mass in the system to derive the mathematical model of such a system. We now demonstrate the process of modelling a system with multi-degrees of freedom.

A three degree-of-freedom mass-spring-damper system. Consider the three degree-of-freedom system shown in Figure 3.20(a), where the masses are constrained to move along the horizontal line. The three external forces $f_1(t)$, $f_2(t)$ and $f_3(t)$ are acting on m_1, m_2 and m_3, respectively. The variables x_1, x_2 and x_3 are the displacements of m_1, m_2 and m_3 from their static equilibrium positions of the associated free system. (As in the single degree-of-freedom system, it is advisable to express the motion of the system from its static equilibrium position).

We draw, without loss of generality, the free body diagram for an instant of time where $0 < x_1 < x_2 < x_3$ and $0 < \dot{x}_1 < \dot{x}_2 < \dot{x}_3$. Then the free body diagram for the mass m_1 gives

$$m_1\ddot{x}_1 = -k_1 x_1 + k_2(x_2 - x_1) + k_5(x_3 - x_1)$$
$$- c_1\dot{x}_1 + c_2(\dot{x}_2 - \dot{x}_1) + c_5(\dot{x}_3 - \dot{x}_1) + f_1(t),$$

(3.87)

or, upon rearranging

$$m_1\ddot{x}_1 + (c_1 + c_2 + c_5)\dot{x}_1 - c_2\dot{x}_2 - c_5\dot{x}_3$$
$$+ (k_1 + k_2 + k_5)x_1 - k_2 x_2 - k_5 x_3 = f_1(t).$$

(3.88)

The free body diagram for m_2 gives

$$m_2\ddot{x}_2 - c_2\dot{x}_1 + (c_2 + c_3)\dot{x}_2 - c_3\dot{x}_3$$
$$- k_2 x_1 + (k_2 + k_3)x_2 - k_3 x_3 = f_2(t).$$

(3.89)

Similarly we find that for m_3 Newton's second law yields

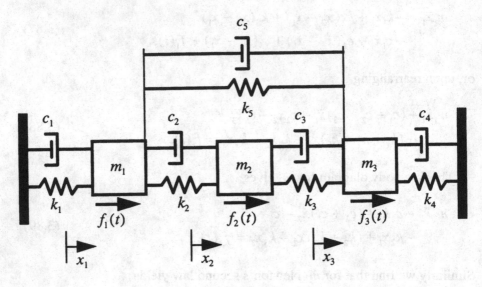

(a) The system

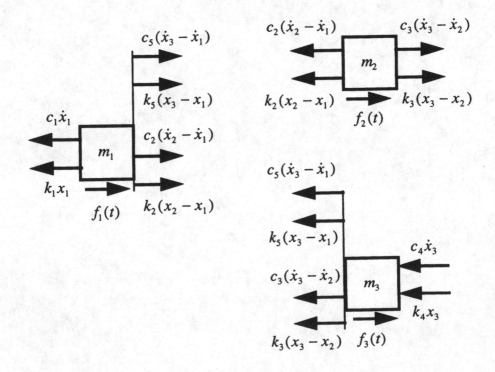

(b) Free body diagrams

Figure 3.20 Three degree-of-freedom system

$$m_3 \ddot{x}_3 - c_5 \dot{x}_1 - c_3 \dot{x}_2 + (c_3 + c_4 + c_5) \dot{x}_3$$
$$- k_5 x_1 - k_3 x_2 + (k_3 + k_4 + k_5) x_3 = f_3(t) \qquad (3.90)$$

Equations (3.88)-(3.90) can be put in the following matrix form

$$
\begin{bmatrix} m_1 & & \\ & m_2 & \\ & & m_3 \end{bmatrix}
\begin{pmatrix} \ddot{x}_1(t) \\ \ddot{x}_2(t) \\ \ddot{x}_3(t) \end{pmatrix}
$$
$$
+ \begin{bmatrix} c_1 + c_2 + c_5 & -c_2 & -c_5 \\ -c_2 & c_2 + c_3 & -c_3 \\ -c_5 & -c_3 & c_3 + c_4 + c_5 \end{bmatrix}
\begin{pmatrix} \dot{x}_1(t) \\ \dot{x}_2(t) \\ \dot{x}_3(t) \end{pmatrix} \qquad (3.91)
$$
$$
+ \begin{bmatrix} k_1 + k_2 + k_5 & -k_2 & -k_5 \\ -k_2 & k_2 + k_3 & -k_3 \\ -k_5 & -k_3 & k_3 + k_4 + k_5 \end{bmatrix}
\begin{pmatrix} x_1(t) \\ x_2(t) \\ x_3(t) \end{pmatrix} = \begin{pmatrix} f_1(t) \\ f_2(t) \\ f_3(t) \end{pmatrix}
$$

where the elements not shown are zeros. Define

$$
\mathbf{x} = \begin{pmatrix} x_1(t) \\ x_2(t) \\ x_3(t) \end{pmatrix}, \mathbf{f} = \begin{pmatrix} f_1(t) \\ f_2(t) \\ f_3(t) \end{pmatrix}, \mathbf{M} = \begin{bmatrix} m_1 & & \\ & m_2 & \\ & & m_3 \end{bmatrix}
$$

$$
\mathbf{C} = \begin{bmatrix} c_1 + c_2 + c_5 & -c_2 & -c_5 \\ -c_2 & c_2 + c_3 & -c_3 \\ -c_5 & -c_3 & c_3 + c_4 + c_5 \end{bmatrix} \qquad (3.92)
$$

$$
\mathbf{K} = \begin{bmatrix} k_1 + k_2 + k_5 & -k_2 & -k_5 \\ -k_2 & k_2 + k_3 & -k_3 \\ -k_5 & -k_3 & k_3 + k_4 + k_5 \end{bmatrix}.
$$

Then the differential equations of motion (3.91) can be symbolically written in the compact form

$$\mathbf{M}\ddot{\mathbf{x}} + \mathbf{C}\dot{\mathbf{x}} + \mathbf{K}\mathbf{x} = \mathbf{f}. \qquad (3.93)$$

The three matrices $\mathbf{M}$, $\mathbf{C}$ and $\mathbf{K}$ are called mass matrix, damping matrix and stiffness matrix, respectively. For the three degrees of freedom analysed here $\mathbf{x}$, $\mathbf{M}$, $\mathbf{C}$, $\mathbf{K}$ and $\mathbf{f}$ are given by (3.92). Equation (3.93), however, holds for a general mass-spring-damper system with many degrees of freedom. In fact it is the most general differential equation of motion governing the motion of a multi-degree-of-freedom *linear time invariant system.*

An *n*-degree-of-freedom mass-spring-damper system. A pattern seems to emerge, rendering the task of determining the equations of motion by inspecting the free body diagrams, as done in the previous section, unnecessary. We first note that the mass matrix is diagonal. Its *i*-th diagonal entry is m_i, ie. the mass associated with the *i*-th degree of freedom. Hence for an *n* degree-of-freedom system the mass matrix is

$$\mathbf{M} = diag\{m_1, m_2, ..., m_n\}. \tag{3.94}$$

Let k_{ij} be the constant of the spring between the *i*-th mass and the *j*-th mass and let k_{i0} be the constant of the spring connecting the *i*-th mass to the ground. Then the *i*-th diagonal element of the stiffness matrix $\mathbf{K}$ is the sum of all the constants of the springs which are attached to the *i*-th mass. The *i-j* element of the stiffness matrix $\mathbf{K}$, for $i \neq j$, is simply minus the constant of the spring connecting the *i*-th mass and the *j*-th mass. For an *n* degree-of-freedom system the stiffness matrix is, therefore,

$$\mathbf{K} = \begin{bmatrix} \displaystyle\sum_{\substack{i=0 \\ i\neq 1}}^{n} k_{1i} & -k_{12} & \cdots & -k_{1n} \\ -k_{21} & \displaystyle\sum_{\substack{i=0 \\ i\neq 2}}^{n} k_{2i} & \cdots & -k_{2n} \\ \cdots & \cdots & \vdots & \cdots \\ -k_{n1} & -k_{n2} & \cdots & \displaystyle\sum_{i=0}^{n-1} k_{ni} \end{bmatrix}. \tag{3.95}$$

The damping matrix $\mathbf{C}$ is similarly constructed with k_{ij} replaced by c_{ij}. Hence

$$
C = \begin{bmatrix} \displaystyle\sum_{\substack{i=0 \\ i \neq 1}}^{n} c_{1i} & -c_{12} & \cdots & -c_{1n} \\[2em] -c_{21} & \displaystyle\sum_{\substack{i=0 \\ i \neq 2}}^{n} c_{2i} & \cdots & -c_{2n} \\[2em] \cdots & \cdots & \vdots & \cdots \\[1em] -c_{n1} & -c_{n2} & \cdots & \displaystyle\sum_{i=0}^{n-1} c_{ni} \end{bmatrix}, \qquad (3.96)
$$

where c_{ij} is the constant of the damper connecting the i-th mass to the j-th mass, c_{i0} is the constant of the damper connecting the i-th mass to the ground.

Obviously, the spring that connects the i-th mass to the j-th mass is the same spring which connects the j-th mass to the i-th mass. We therefore have

$$
k_{ij} = k_{ji}, \qquad (3.97)
$$

and we thus conclude that by (3.95) the stiffness matrix K is symmetric. Using an identical argument we find that

$$
c_{ij} = c_{ji} \qquad (3.98)
$$

and hence the damping matrix C is symmetric as well. It should be emphasised that this approach holds only if x_i is the displacement of the i-th degree-of-freedom measured from the static equilibrium position of the associated free system. For such a case the spring and damper forces are symmetric about the system's configuration. If the displacement is measured from a different reference point we cannot use (3.94)-(3.96). In such a case the matrices of the system are not necessarily symmetric.

Test Example 3.1.
Determine the equations of motion for the five degree-of-freedom system shown in Figure 3.21 by inspection.

Solution.
The matrix differential equation of motion is

$$
M\ddot{x} + C\dot{x} + Kx = o
$$

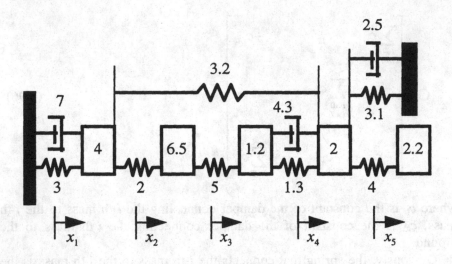

Figure 3.21 A five degree-of-freedom system

where

$$\mathbf{M} = diag\{4,\ 6.5,\ 1.2,\ 2,\ 2.2\},$$

$$\mathbf{C} = \begin{bmatrix} 7 & 0 & 0 & 0 & 0 \\ 0 & 0 & 0 & 0 & 0 \\ 0 & 0 & 4.3 & -4.3 & 0 \\ 0 & 0 & -4.3 & 6.8 & 0 \\ 0 & 0 & 0 & 0 & 0 \end{bmatrix},\ \mathbf{K} = \begin{bmatrix} 8.2 & -2 & 0 & -3.2 & 0 \\ -2 & 7 & -5 & 0 & 0 \\ 0 & -5 & 6.3 & -1.3 & 0 \\ -3.2 & 0 & -1.3 & 11.6 & -4 \\ 0 & 0 & 0 & -4 & 4 \end{bmatrix}.$$

The double pendulum. Newton's second law, or the moment-rotation relation (3.10), could be applied to determine the linearised differential equations governing the small oscillations about equilibrium of the double pendulum shown in Figure 3.22. We begin by analysing the (simpler) force-acceleration relation for m_2. It follows from the free body diagram shown in Figure 3.23(b) that along the line B-B, (perpendicular to T_2), there is an action of only one force component. Newton's second law is thus applied in the direction B-B and for small angle θ_2 to give

$$m_2\left(L_1\ddot{\theta}_1 + L_2\ddot{\theta}_2\right) = -m_2 g\theta_2, \tag{3.99}$$

or

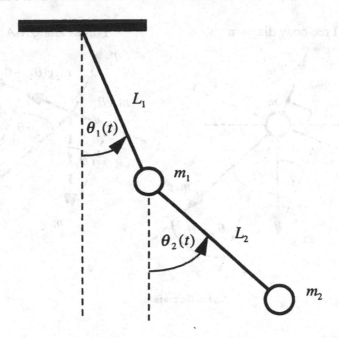

Figure 3.22. The double pendulum

$$L_1\ddot{\theta}_1 + L_2\ddot{\theta}_2 + g\theta_2 = 0. \tag{3.100}$$

The radial acceleration in the direction perpendicular to *B-B* is of order $\dot{\theta}^2$ (see equation 3.3c). We therefore neglect this acceleration when analysing the small vibration and obtain

$$T_2 = mg. \tag{3.101}$$

The free body diagram for m_1 is shown in Figure 3.23(a). The component of the tension force T_1 vanishes in the direction *A-A*. Newton's second law applied in this direction gives

$$m_1\left(L_1\ddot{\theta}_1\right) = -m_1 g\theta_1 + m_2 g(\theta_2 - \theta_1), \tag{3.102}$$

or

$$m_1 L_1\ddot{\theta}_1 + (m_1 + m_2)g\theta_1 - m_2 g\theta_2 = 0. \tag{3.103}$$

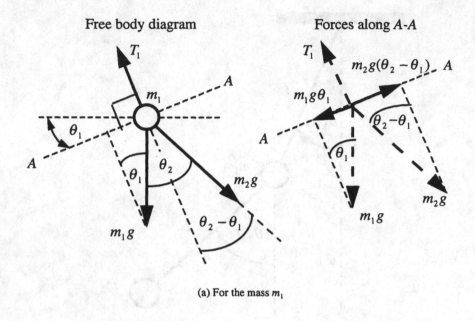

(a) For the mass m_1

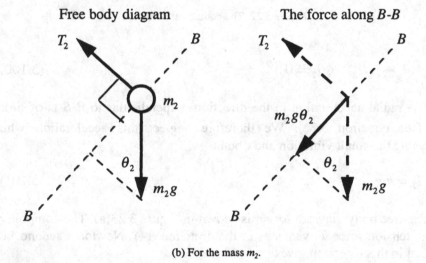

(b) For the mass m_2.

Figure 3.23 Free body diagrams for the double pendulum

The differential equations of motion (3.103) and (3.100) can be written in matrix form as

$$
\begin{bmatrix} m_1 L_1 & 0 \\ L_1 & L_2 \end{bmatrix}
\begin{pmatrix} \ddot{\theta}_1 \\ \ddot{\theta}_2 \end{pmatrix}
+
\begin{bmatrix} (m_1+m_2)g & -m_2 g \\ 0 & g \end{bmatrix}
\begin{pmatrix} \theta_1 \\ \theta_2 \end{pmatrix}
=
\begin{pmatrix} 0 \\ 0 \end{pmatrix},
\qquad (3.104)
$$

or symbolically

$$\mathbf{M}\ddot{\theta} + \mathbf{K}\theta = \mathbf{o},$$ (3.105)

where

$$\mathbf{M} = \begin{bmatrix} m_1 L_1 & 0 \\ L_1 & L_2 \end{bmatrix}, \ \mathbf{K} = \begin{bmatrix} (m_1 + m_2)g & -m_2 g \\ 0 & g \end{bmatrix} \text{ and } \theta = \begin{pmatrix} \theta_1 \\ \theta_2 \end{pmatrix}.$$

We see that here neither $\mathbf{M}$ nor $\mathbf{K}$ are symmetric. This may be expected as the displacement of m_2 is not expressed from the static equilibrium position.

3.6 The Response of Multi-Degree-of-Freedom Systems

Symmetric undamped system. Consider an n degree-of-freedom undamped (ie. $\mathbf{C}=\mathbf{0}$) system. Its matrix equation of motion (3.93) reduces to

$$\mathbf{M}\ddot{\mathbf{x}} + \mathbf{K}\mathbf{x} = \mathbf{f}(t),$$ (3.106)

where $\mathbf{M}$ and $\mathbf{K}$ are $n{\times}n$ matrices. Suppose further that the motion is measured from the static equilibrium position, so that $\mathbf{M}$ and $\mathbf{K}$ are also symmetric. In this section we focus our attention on how to determine the solution of (3.106), ie. the determination of $\mathbf{x}(t)$, assuming that $\mathbf{M}$, $\mathbf{K}$ and $\mathbf{f}(t)$ are given. To sense the difficulty arising in solving such a problem, consider as a simple example the two degree-of-freedom system modelled by

$$\mathbf{M} = \begin{bmatrix} 1 & \\ & 1 \end{bmatrix}, \ \mathbf{K} = \begin{bmatrix} 2 & -1 \\ -1 & 1 \end{bmatrix} \text{ and } \mathbf{f} = \begin{pmatrix} 0 \\ 0 \end{pmatrix}.$$

The differential equations of motion for this system are therefore

$$\begin{cases} \ddot{x}_1 + 2x_1 - x_2 = 0 \\ \ddot{x}_2 - x_1 + x_2 = 0 \end{cases}.$$ (3.107)

We cannot solve the first equation of (3.107) alone for $x_1(t)$, since $x_2(t)$ is unknown. Similarly, we cannot solve the second equation of (3.107) alone for $x_2(t)$, since $x_1(t)$ is unknown. The difficulty is that these equations are coupled. We thus need to show how a system of n coupled equations can be decoupled into n equations of second order. Each of the decoupled equations can then be solved by using the methods described in section 3.3.

Before showing how to decouple the equations we need to do some background work. Consider the free system

$$\mathbf{M\ddot{x} + Kx = o} \tag{3.108}$$

associated with (3.106). Let us try a solution of the form

$$\mathbf{x}(t) = \phi \sin \omega_n t \tag{3.109}$$

where ϕ is a constant vector (ie. independent of the time t) of the same dimension as $\mathbf{x}(t)$. The physical meaning of (3.109) is that each mass in the system vibrates with the same frequency ω_n, but with possibly different amplitude of vibration. Indeed at this stage we do not know that this is the case. We only try a solution of the form (3.109) to equation (3.108). But if (3.109) is a valid solution to (3.108) then clearly ω_n is a natural frequency of the oscillation of the undamped free system. Note here that ω_n is not necessarily equal to $\sqrt{k/m}$ as in the previous sections dealing with the single degree of freedom system.

Substituting (3.109) in (3.108) gives

$$-\omega_n^2 \mathbf{M}\phi + \mathbf{K}\phi = \mathbf{o}. \tag{3.110}$$

Denoting

$$\lambda = \omega_n^2 \tag{3.111}$$

equation (3.110) can be written as

$$(\mathbf{K} - \lambda \mathbf{M})\phi = \mathbf{o}. \tag{3.112}$$

Equation (3.112) is called the *generalised eigenvalue problem*. It has one obvious solution $\phi = \mathbf{o}$. We are not interested in this trivial solution that actually only tells us that under certain conditions the system (3.108) can be at rest. Noting that $\mathbf{K} - \lambda \mathbf{M}$ is some matrix, we invoke the well known result of linear algebra, stating that the system of equation $\mathbf{Ax = b}$ has a unique solution if and only if $\mathbf{A}$ is nonsingular. We thus conclude that (3.112) has other, nontrivial solutions provided that $\mathbf{K} - \lambda \mathbf{M}$ is singular, ie.

$$\det(\mathbf{K} - \lambda \mathbf{M}) = 0 . \tag{3.113}$$

For mass and stiffness matrices associated with vibratory systems $\det(\mathbf{K} - \lambda \mathbf{M})$ is always a polynomial in λ of degree n. Hence (3.113) has n possible roots. Denote the roots of (3.113) by λ_i, $i=1,2,...,n$. There are also n eigenvectors ϕ_i associated with λ_i, $i=1,2,...,n$. They are the solutions of

$$(\mathbf{K} - \lambda_i \mathbf{M})\phi_i = \mathbf{o}, \qquad i = 1,2,...,n . \tag{3.114}$$

The eigenvector is determined only up to a scale factor. If ϕ_i is an eigenvector which solves (3.114) then $\alpha \phi_i$, with α some scalar, is another eigenvector of the same problem. The pair, an eigenvalue λ_i and its associated eigenvector ϕ_i, is called an *eigenpair*.

We now make the additional assumption that the system (3.114) has distinct eigenvalues, ie. $\lambda_i \neq \lambda_j$ for $i \neq j$. Denote the *spectral matrix* by

$$\Lambda = \begin{bmatrix} \lambda_1 & & & \\ & \lambda_2 & & \\ & & \ddots & \\ & & & \lambda_n \end{bmatrix} . \tag{3.115}$$

The order in which the eigenvalues appears in (3.115) can be chosen arbitrarily. To avoid arbitrariness, however, it is useful to order the eigenvalues in increasing order, ie. $\lambda_1 < \lambda_2 <...< \lambda_n$. We also define a *modal matrix* by

$$\Phi = [\phi_1 \mid \phi_2 \mid ... \mid \phi_n] . \tag{3.116}$$

Note that the ordering of the diagonal entries of Λ determines the ordering of the eigenvectors in Φ. With this definition the n equations of (3.114) can be written in the following matrix form

$$\mathbf{K}\Phi = \mathbf{M}\Phi\Lambda . \tag{3.117}$$

Equation (3.117) is one of the central equations in vibration analysis.

We now establish the following results.

Theorem 3.1

Let $\mathbf{M}, \mathbf{K}$ be symmetric matrices where $\mathbf{M}$ is positive definite. Suppose that the eigenvalues of $\mathbf{K} - \lambda \mathbf{M}$ are distinct, ie. $\lambda_i \neq \lambda_j$ for $i \neq j$. Then

(*i*) The products $\mathbf{\Phi}^T \mathbf{K} \mathbf{\Phi}$ and $\mathbf{\Phi}^T \mathbf{M} \mathbf{\Phi}$ are both diagonal.
(*ii*) Furthermore, let

$$\mathbf{\Phi}^T \mathbf{K} \mathbf{\Phi} = \begin{bmatrix} \alpha_1 & & & \\ & \alpha_2 & & \\ & & \ddots & \\ & & & \alpha_n \end{bmatrix} \tag{3.118}$$

and

$$\mathbf{\Phi}^T \mathbf{M} \mathbf{\Phi} = \begin{bmatrix} \beta_1 & & & \\ & \beta_2 & & \\ & & \ddots & \\ & & & \beta_n \end{bmatrix}. \tag{3.119}$$

Then

$$\lambda_i = \frac{\alpha_i}{\beta_i}, \qquad i = 1, 2, \ldots, n. \tag{3.120}$$

Proof.

We will use the following simple facts:
    (*a*) A matrix $\mathbf{A}$ is symmetric if and only if $\mathbf{A} = \mathbf{A}^T$, and,
    (*b*) for two matrices $\mathbf{A}$ and $\mathbf{B}$ of appropriate dimensions we have $(\mathbf{AB})^T = \mathbf{B}^T \mathbf{A}^T$.

It follows immediately from (*b*) that for three matrices $\mathbf{A}$, $\mathbf{B}$ and $\mathbf{C}$ we have $(\mathbf{ABC})^T = \mathbf{C}^T (\mathbf{AB})^T = \mathbf{C}^T \mathbf{B}^T \mathbf{A}^T$. Hence, since $\mathbf{K}$ is symmetric, $\mathbf{K} = \mathbf{K}^T$, and the product $(\mathbf{\Phi}^T \mathbf{K} \mathbf{\Phi})^T = \mathbf{\Phi}^T \mathbf{K}^T \mathbf{\Phi} = \mathbf{\Phi}^T \mathbf{K} \mathbf{\Phi}$ must be symmetric. By an identical argument we see that the product $\mathbf{\Phi}^T \mathbf{M} \mathbf{\Phi}$ is also symmetric.

Denote

$$\mathbf{F} = \mathbf{\Phi}^T \mathbf{K} \mathbf{\Phi} \tag{3.121}$$

and

$$\mathbf{G} = \mathbf{\Phi}^T \mathbf{M} \mathbf{\Phi}. \tag{3.122}$$

Then $\mathbf{F}$ and $\mathbf{G}$ are symmetric. Let

$$\mathbf{F} = \begin{bmatrix} \alpha_1 & f_{12} & \cdots & f_{1n} \\ f_{21} & \alpha_2 & \cdots & f_{2n} \\ \cdots & \cdots & \vdots & \cdots \\ f_{n1} & f_{n2} & \cdots & \alpha_n \end{bmatrix} \text{ and } \mathbf{G} = \begin{bmatrix} \beta_1 & g_{12} & \cdots & g_{1n} \\ g_{21} & \beta_2 & \cdots & g_{2n} \\ \cdots & \cdots & \vdots & \cdots \\ g_{n1} & g_{n2} & \cdots & \beta_n \end{bmatrix}.$$

Then because $\mathbf{F}$ and $\mathbf{G}$ are symmetric we have

$$f_{ij} = f_{ji}, \quad i \neq j \tag{3.123}$$

and

$$g_{ij} = g_{ji}, \quad i \neq j. \tag{3.124}$$

Pre-multiplying (3.117) by $\mathbf{\Phi}^T$ gives

$$\mathbf{\Phi}^T \mathbf{K} \mathbf{\Phi} = \mathbf{\Phi}^T \mathbf{M} \mathbf{\Phi} \mathbf{\Lambda}. \tag{3.125}$$

Hence, by (3.121) and (3.122)

$$\mathbf{F} = \mathbf{G} \mathbf{\Lambda}. \tag{3.126}$$

Componentwise, (3.126) has the form

$$\begin{bmatrix} \alpha_1 & f_{12} & \cdots & f_{1n} \\ f_{21} & \alpha_2 & \cdots & f_{2n} \\ \cdots & \cdots & \vdots & \cdots \\ f_{n1} & f_{n2} & \cdots & \alpha_n \end{bmatrix} = \begin{bmatrix} \lambda_1 \beta_1 & \lambda_2 g_{12} & \cdots & \lambda_n g_{1n} \\ \lambda_1 g_{21} & \lambda_2 \beta_2 & \cdots & \lambda_n g_{2n} \\ \cdots & \cdots & \vdots & \cdots \\ \lambda_1 g_{n1} & \lambda_2 g_{n2} & \cdots & \lambda_n \beta_n \end{bmatrix}. \tag{3.127}$$

The 1-2 element of (3.127) gives

$$f_{12} = \lambda_2 g_{12} \tag{3.128}$$

and the 2-1 element of (3.127) gives

$$f_{21} = \lambda_1 g_{21}.$$

(3.129)

But by (3.123) $f_{21}=f_{12}$ and by (3.124) $g_{21}=g_{12}$. Hence (3.129) can be rewritten as

$$f_{12} = \lambda_1 g_{12}.$$

(3.130)

Subtracting (3.130) from (3.128) yields

$$0 = (\lambda_2 - \lambda_1) g_{12}.$$

(3.131)

Hence by the assumption that the eigenvalues are distinct, $\lambda_1 \neq \lambda_2$ and we obtain $g_{12}=0$. Then (3.124), (3.128) and (3.129) yield $g_{21}=f_{12}=f_{21}=0$.

By repeating this process and considering both the i-j and j-i elements of (3.127) we see that all the off-diagonal elements of $\mathbf{F}$ and $\mathbf{G}$ vanish. Hence $\mathbf{F} = \Phi^T \mathbf{K} \Phi$ and $\mathbf{G} = \Phi^T \mathbf{M} \Phi$ are diagonal matrices and (*i*) is proved.

The i-i element of (3.127) gives

$$\alpha_i = \lambda_i \beta_i.$$

(3.132)

Hence

$$\lambda_i = \frac{\alpha_i}{\beta_i}$$

(1.133)

which completes the proof of the theorem.   ∎

We can now show how to decouple the equations of motion (3.106). Pre-multiplying (3.106) by Φ^T gives

$$\Phi^T \mathbf{M} \ddot{\mathbf{x}} + \Phi^T \mathbf{K} \mathbf{x} = \Phi^T \mathbf{f}(t).$$

(3.134)

This equation can be written as

$$\Phi^T \mathbf{M} \Phi \Phi^{-1} \ddot{\mathbf{x}} + \Phi^T \mathbf{K} \Phi \Phi^{-1} \mathbf{x} = \Phi^T \mathbf{f}(t).$$

(3.135)

Denote

$$\mathbf{z} = \Phi^{-1} \mathbf{x},$$

(3.136)

then (3.135) becomes

$$\Phi^T M \Phi \ddot{z} + \Phi^T K \Phi z = \Phi^T f(t) \,. \tag{3.137}$$

We now invoke (3.118) and (3.119) and obtain

$$\begin{bmatrix} \beta_1 & & \\ & \ddots & \\ & & \beta_n \end{bmatrix} \begin{pmatrix} \ddot{z}_1 \\ \vdots \\ \ddot{z}_n \end{pmatrix} + \begin{bmatrix} \alpha_1 & & \\ & \ddots & \\ & & \alpha_n \end{bmatrix} \begin{pmatrix} z_1 \\ \vdots \\ z_n \end{pmatrix} = \begin{pmatrix} \sum_{i=1}^n \phi_{i1} f_i \\ \vdots \\ \sum_{i=1}^n \phi_{in} f_i \end{pmatrix} \tag{3.138}$$

This is a decoupled system of equations. It can be written in the following scalar form

$$\beta_j \ddot{z}_j + \alpha_j z_j = \sum_{i=1}^n \phi_{ij} f_i \,, \qquad j = 1,2,\ldots,n \,. \tag{3.139}$$

Equation (3.139) describes the motion of a single degree of freedom system. Denoting

$$\hat{f}_j = \sum_{i=1}^n \phi_{ij} f_i \,, \qquad j = 1,2,\ldots,n \,, \tag{3.140}$$

the general solution of (3.139), when $\lambda_j > 0$, is given by

$$\begin{aligned} z_j(t) &= a_j \sin \sqrt{\lambda_j} \, t + b_j \cos \sqrt{\lambda_j} \, t \\ &\quad + \int_0^t \hat{f}_j(\tau) h_j(t-\tau) d\tau \end{aligned} \tag{3.141}$$

where (using (3.59) with $\zeta = 0$)

$$h_j(t-\tau) = \begin{cases} 0 \,, & t < \tau \\ \dfrac{1}{\beta_j \sqrt{\lambda_j}} \sin\left(\sqrt{\lambda_j}(t-\tau)\right) \,, & t > \tau \end{cases} \tag{3.142}$$

By (3.136), the initial conditions in terms of z are

$$\mathbf{z}(0) = \Phi^{-1}\mathbf{x}(0) \tag{3.143}$$

and

$$\dot{\mathbf{z}}(0) = \Phi^{-1}\dot{\mathbf{x}}(0), \tag{3.144}$$

where

$$\mathbf{z} = \left(z_1, z_2, ..., z_n\right)^T. \tag{3.145}$$

The coefficients a_j, b_j for $j=1,2,...,n$, can thus be determined by using (3.141) together with (3.143) and (3.144). Alternatively equation (3.139) together with the initial conditions (3.143) and (3.144) can be solved by the Laplace transform method. This determines the vector $\mathbf{z}(t)$.

The response $\mathbf{x}(t)$ is finally determined by (3.136), ie.

$$\mathbf{x}(t) = \Phi\mathbf{z}(t). \tag{3.146}$$

Free non-symmetric undamped system. The motion of an n degree-of-freedom free undamped system is governed by the system of differential equations

$$\mathbf{M}\ddot{\mathbf{x}} + \mathbf{K}\mathbf{x} = \mathbf{0} \tag{3.147}$$

and the initial conditions

$$\begin{cases} \mathbf{x}(0) = \mathbf{x}_0 \\ \dot{\mathbf{x}}(0) = \mathbf{v}_0 \end{cases}. \tag{3.148}$$

The response $\mathbf{x}(t)$ in this case can be obtained without decoupling the equations of motion (3.147).

Let λ_i and ϕ_i, $i=1,2,...,n$, be the eigenvalues and eigenvectors of

$$(\mathbf{K} - \lambda_i\mathbf{M})\phi_i = \mathbf{0}. \tag{3.149}$$

Then it is easily shown by direct substitution that if $\lambda_i \neq 0$ then the vector

$$a_i\phi_i \sin\sqrt{\lambda_i}\,t \tag{3.150}$$

satisfies the differential equation (3.147), where $a_i \neq 0$ is an arbitrary constant. Similarly, for an arbitrary constant $b_i \neq 0$, the vector

$$b_i \phi_i \cos \sqrt{\lambda_i} t \tag{3.151}$$

satisfies (3.147) as well.

Hence by the linearity of the problem we find that the sum of these solutions is also a solution to the same problem, ie.

$$\mathbf{x}(t) = \sum_{i=1}^{n} \left(a_i \phi_i \sin \sqrt{\lambda_i} t + b_i \phi_i \cos \sqrt{\lambda_i} t \right) \tag{3.152}$$

satisfies the differential equation (3.147). But $\mathbf{x}(t)$ in (3.152) is expressed in terms of $2n$ arbitrary constants and it is therefore the general solution of (3.147). The coefficients a_i and b_i are determined by the initial conditions (3.148) as follows. Using (3.152) the initial displacement gives

$$\mathbf{x}(0) = \sum_{i=1}^{n} b_i \phi_i = \mathbf{x}_0 . \tag{3.153}$$

We may express (3.153) in matrix form as

$$\Phi \mathbf{b} = \mathbf{x}_0 , \tag{3.154}$$

where the modal matrix Φ is as in (3.116) and $\mathbf{b} = (b_1, b_2, ..., b_n)^T$. We thus have

$$\mathbf{b} = \Phi^{-1} \mathbf{x}_0 . \tag{3.155}$$

The initial velocity condition, substituted in (3.152), gives

$$\dot{\mathbf{x}}(0) = \sum_{i=1}^{n} a_i \sqrt{\lambda_i} \phi_i = \mathbf{v}_0 , \tag{3.156}$$

which can be written in matrix form as

$$\Phi \Lambda^{1/2} \mathbf{a} = \mathbf{v}_0 , \tag{3.157}$$

where

$$\Lambda^{1/2} = diag\{\sqrt{\lambda_1}, \sqrt{\lambda_2}, ..., \sqrt{\lambda_n}\}$$

and

$$\mathbf{a} = (a_1, a_2, ..., a_n)^T.$$

We thus have

$$\mathbf{a} = \Lambda^{-1/2} \Phi^{-1} \mathbf{v}_0. \tag{3.158}$$

In summary, the general solution of (3.147) is given by (3.152), where the coefficients a_i and b_i, $i=1,2,...,n$, are determined by (3.158) and (3.155), respectively. The following example demonstrates this result.

Test Example 3.2

Determine the response of the system modelled by (3.147) where

$$\mathbf{M} = \begin{bmatrix} 1 & \\ & 1 \end{bmatrix}, \mathbf{K} = \begin{bmatrix} 3 & -2 \\ -2 & 3 \end{bmatrix}, \mathbf{x}(0) = \begin{pmatrix} 1 \\ -1 \end{pmatrix} \text{ and } \dot{\mathbf{x}}(0) = \begin{pmatrix} 1 \\ 0 \end{pmatrix}.$$

Solution

The spectral and modal matrices for this system are

$$\Lambda = \begin{bmatrix} 1 & \\ & 5 \end{bmatrix} \text{ and } \Phi = \begin{bmatrix} 1 & 1 \\ 1 & -1 \end{bmatrix}.$$

Hence with

$$\mathbf{x}_0 = \begin{pmatrix} 1 \\ -1 \end{pmatrix}, \quad \mathbf{v}_0 = \begin{pmatrix} 1 \\ 0 \end{pmatrix}$$

we have by (3.158)

$$\begin{pmatrix} a_1 \\ a_2 \end{pmatrix} = \begin{bmatrix} 1 & \\ & \frac{1}{\sqrt{5}} \end{bmatrix} \begin{bmatrix} 0.5 & 0.5 \\ 0.5 & -0.5 \end{bmatrix} \begin{pmatrix} 1 \\ 0 \end{pmatrix} = \begin{pmatrix} 0.5 \\ \frac{1}{2\sqrt{5}} \end{pmatrix}$$

and by (3.155)

$$\begin{pmatrix} b_1 \\ b_2 \end{pmatrix} = \begin{bmatrix} 0.5 & 0.5 \\ 0.5 & -0.5 \end{bmatrix} \begin{pmatrix} 1 \\ -1 \end{pmatrix} = \begin{pmatrix} 0 \\ 1 \end{pmatrix}.$$

The response of the system is thus

$$\begin{pmatrix} x_1(t) \\ x_2(t) \end{pmatrix} = \frac{1}{2}\begin{pmatrix} 1 \\ 1 \end{pmatrix}\sin t + \frac{1}{2\sqrt{5}}\begin{pmatrix} 1 \\ -1 \end{pmatrix}\sin\sqrt{5}t + \begin{pmatrix} 1 \\ -1 \end{pmatrix}\cos\sqrt{5}t$$

or

$$\begin{cases} x_1(t) = \dfrac{1}{2}\sin t + \dfrac{1}{2\sqrt{5}}\sin\sqrt{5}t + \cos\sqrt{5}t \\ x_2(t) = \dfrac{1}{2}\sin t - \dfrac{1}{2\sqrt{5}}\sin\sqrt{5}t - \cos\sqrt{5}t \end{cases}.$$

We note that if $\lambda_1 = 0$ then $a_1\phi_1 \sin\sqrt{\lambda_1}t = 0$ and there are only $2n$-1 arbitrary constants in (3.152). We thus conclude that in this case (3.152) is not the general solution of (3.147). However

$$\mathbf{x}(t) = a_1 t \phi_1, \tag{3.159}$$

for an arbitrary scalar a_1, does solve (3.147) in this case, as will now be shown. If $\lambda_1 = 0$ then (3.149) implies that

$$\mathbf{K}\phi_1 = \mathbf{o}. \tag{3.160}$$

Substituting (3.159) in (3.147) and using (3.160) we obtain

$$\mathbf{M}\frac{d^2}{dt^2}(a_1 t\phi_1) + a_1 t \mathbf{K}\phi_1 = \mathbf{o}$$

which proves the above statement.

It thus follows that when $\lambda_1 = 0$, the general solution of (3.147) is given by

$$\mathbf{x}(t) = a_1 t \phi_1 + b_1 \phi_1 + \sum_{i=2}^{n} \left(a_i \phi_i \sin \sqrt{\lambda_i} t + b_i \phi_i \cos \sqrt{\lambda_i} t \right). \qquad (3.161)$$

We see that there is a non-oscillatory mode of motion, namely $(a_1 t + b_1) \phi_1$, in which the system moves from its static equilibrium position without bound as t increases. This mode of motion is called a rigid body mode of motion. Clearly the system can have such a mode of motion only if there is no spring connections between the system and the ground. In this case $\lambda_1 = 0$ and the response of the system is determined by (3.161).

The $2n$ coefficients in (3.161) are determined by the initial conditions. The b_i, $i=1,2,...,n$, can be obtained by using (3.155). The a_i, $i=1,2,...,n$, however, cannot be obtained by (3.158), as Λ is singular and $\Lambda^{-1/2}$ does not exist. We may obtain these coefficients directly from the initial conditions as demonstrated by the following example.

Test Example 3.3
Determine the response to (3.147) where

$$\mathbf{M} = \begin{bmatrix} 1 & & \\ & 2 & \\ & & 1 \end{bmatrix}, \ \mathbf{K} = \begin{bmatrix} 5 & -5 & 0 \\ -5 & 10 & -5 \\ 0 & -5 & 5 \end{bmatrix}, \ \mathbf{x}(0) = \begin{pmatrix} x_0 \\ 0 \\ 0 \end{pmatrix}, \ \dot{\mathbf{x}}(0) = \begin{pmatrix} v_0 \\ 0 \\ 0 \end{pmatrix}.$$

Solution

The spectral and modal matrices for this system are

$$\Lambda = \begin{bmatrix} 0 & & \\ & 5 & \\ & & 10 \end{bmatrix} \text{ and } \Phi = \begin{bmatrix} 1 & 1 & 1 \\ 1 & 0 & -1 \\ 1 & -1 & 1 \end{bmatrix}.$$

So here $\lambda_1 = 0$ and the general solution of this problem, given by (3.161), is

$$\mathbf{x}(t) = a_1 t \begin{pmatrix} 1 \\ 1 \\ 1 \end{pmatrix} + b_1 \begin{pmatrix} 1 \\ 1 \\ 1 \end{pmatrix} + a_2 \begin{pmatrix} 1 \\ 0 \\ -1 \end{pmatrix} \sin\sqrt{5}t + b_2 \begin{pmatrix} 1 \\ 0 \\ -1 \end{pmatrix} \cos\sqrt{5}t$$

$$+ a_3 \begin{pmatrix} 1 \\ -1 \\ 1 \end{pmatrix} \sin\sqrt{10}t + b_3 \begin{pmatrix} 1 \\ -1 \\ 1 \end{pmatrix} \cos\sqrt{10}t$$

(3.162)

Noting that

$$\Phi^{-1} = \begin{bmatrix} 0.25 & 0.5 & 0.25 \\ 0.5 & 0 & -0.5 \\ 0.25 & -0.5 & 0.25 \end{bmatrix}$$

we have by (3.155)

$$\begin{pmatrix} b_1 \\ b_2 \\ b_3 \end{pmatrix} = \begin{bmatrix} 0.25 & 0.5 & 0.25 \\ 0.5 & 0 & -0.5 \\ 0.25 & -0.5 & 0.25 \end{bmatrix} \begin{pmatrix} x_0 \\ 0 \\ 0 \end{pmatrix} = \begin{pmatrix} 0.25x_0 \\ 0.5x_0 \\ 0.25x_0 \end{pmatrix}.$$

Differentiating (3.162) with respect to t gives

$$\dot{\mathbf{x}}(t) = a_1 \begin{pmatrix} 1 \\ 1 \\ 1 \end{pmatrix} + a_2\sqrt{5} \begin{pmatrix} 1 \\ 0 \\ -1 \end{pmatrix} \cos\sqrt{5}t - b_2\sqrt{5} \begin{pmatrix} 1 \\ 0 \\ -1 \end{pmatrix} \sin\sqrt{5}t$$

$$+ a_3\sqrt{10} \begin{pmatrix} 1 \\ -1 \\ 1 \end{pmatrix} \cos\sqrt{10}t - b_3\sqrt{10} \begin{pmatrix} 1 \\ -1 \\ 1 \end{pmatrix} \sin\sqrt{10}t.$$

Hence at $t=0$ we have the following system of equations

$$\begin{cases} v_0 = a_1 + \sqrt{5}a_2 + \sqrt{10}a_3 \\ 0 = a_1 \qquad\qquad - \sqrt{10}a_3 , \\ 0 = a_1 - \sqrt{5}a_2 + \sqrt{10}a_3 \end{cases}$$

which has the solution

$$a_1 = \frac{v_0}{4}, \ a_2 = \frac{v_0}{2\sqrt{5}} \ \text{and} \ a_3 = \frac{v_0}{4\sqrt{10}}.$$

Substituting a_i and b_i, $i=1,2,3$, in (3.162) gives the system response

$$\mathbf{x}(t) = \frac{v_0 t}{4}\begin{pmatrix} 1 \\ 1 \\ 1 \end{pmatrix} + \frac{x_0}{4}\begin{pmatrix} 1 \\ 1 \\ 1 \end{pmatrix} + \frac{v_0}{2\sqrt{5}}\begin{pmatrix} 1 \\ 0 \\ -1 \end{pmatrix}\sin\sqrt{5}t + \frac{x_0}{2}\begin{pmatrix} 1 \\ 0 \\ -1 \end{pmatrix}\cos\sqrt{5}t$$

$$+ \frac{v_0}{4\sqrt{10}}\begin{pmatrix} 1 \\ -1 \\ 1 \end{pmatrix}\sin\sqrt{10}t + \frac{x_0}{4}\begin{pmatrix} 1 \\ -1 \\ 1 \end{pmatrix}\cos\sqrt{10}t$$

We see that the system moves to the left according to its rigid body mode of motion $(v_0 t + x_0)/4$, ie. with constant velocity $0.25v_0$. There are two additional components of vibration, one with frequency $\sqrt{5}$, the other with frequency $\sqrt{10}$.

We note that in the above analysis no assumption regarding the symmetry of $\mathbf{M}$ and $\mathbf{K}$ has been made.

Harmonic excitation of undamped systems. If an undamped system is excited by the force $\mathbf{f}(t) = \mathbf{g}\sin\omega t$, where $\mathbf{g}$ is a constant vector, then the matrix differential equation governing the vibrations is

$$\mathbf{M\ddot{x}}(t) + \mathbf{Kx}(t) = \mathbf{g}\sin\omega t. \tag{3.163}$$

We try a particular solution of the form

$$\mathbf{x_p} = \mathbf{u}\sin\omega t, \tag{3.164}$$

where $\mathbf{u}$ is a constant vector. Upon substituting $\mathbf{x_p}$ for $\mathbf{x}$ in (3.163) we obtain

$$-\omega^2\mathbf{Mu}\sin\omega t + \mathbf{Ku}\sin\omega t = \mathbf{g}\sin\omega t, \tag{3.165}$$

or

$$\left(K - \omega^2 M\right) u = g. \tag{3.166}$$

So

$$u = \left(K - \omega^2 M\right)^{-1} g, \tag{3.167}$$

provided that $K - \omega^2 M$ is invertible. By the definition of λ_i the matrix $K - \omega^2 M$ is invertible if $\omega^2 \neq \lambda_i$, for $i=1,2,...,n$, and singular if $\omega^2 = \lambda_i$. In the case where $\omega = \sqrt{\lambda_j}$ there is generally no vector u which satisfies (3.166), and therefore there is no particular solution x_p of the form (3.164). The system is in resonance when the exciting frequency ω equals one of natural frequencies $\sqrt{\lambda_j}$ of the system and its response increases without bound.

The general solution of (3.163) is obtained by adding the particular solution x_P to the general solution x_H of the associated homogeneous problem, as determined by either (3.152) or (3.161).

3.7 Distributed Parameter Systems

The vibrating string. Consider a uniform string, of mass per unit length ρ, stretched under tension T, between two supports distance L apart, as shown in Figure 3.24(a). With the assumption of small vibrations about equilibrium, the tension T in the spring may be considered constant, ie. independent of the time t and the position x. The free body diagram for an element of the string with infinitesimal length dx, lying originally in the interval $[x,x+dx]$, is shown in Figure 3.24(b).

The mass of the element is ρdx, its acceleration is $\partial^2 u / \partial t^2$, and Newton's second law, applied in the u direction, gives

$$\rho \, dx \frac{\partial^2 u}{\partial t^2} = T\left(\theta + \frac{\partial \theta}{\partial x} dx\right) - T\theta, \tag{3.168}$$

which simplifies to

$$\rho \frac{\partial^2 u}{\partial t^2} = T \frac{\partial \theta}{\partial x}. \tag{3.169}$$

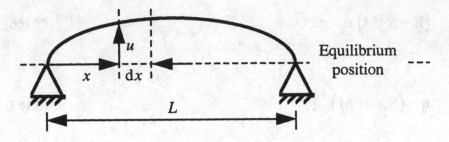

(a) The system

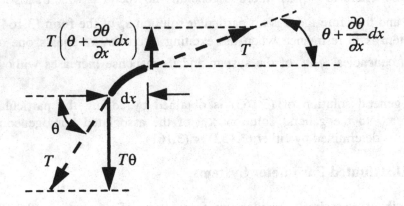

(b) Free body diagram

Figure 3.24 The vibrating string

But, by definition the slope $\theta(x,t)$ is

$$\theta = \frac{\partial u}{\partial x}. \tag{3.170}$$

Hence, substituting (3.170) in (3.169) gives

$$\rho\frac{\partial^2 u}{\partial t^2} = T\frac{\partial^2 u}{\partial x^2}, \tag{3.171}$$

or

$$\frac{\partial^2 u}{\partial t^2} = c^2 \frac{\partial^2 u}{\partial x^2},$$ (3.172)

where

$$c = \sqrt{T/\rho}.$$ (3.173)

The parameter c is known as the speed propagation of waves in the string.

The dynamics of the string are determined by the differential equation (3.172), the boundary conditions

$$\begin{cases} u(0,t) = 0 \\ u(L,t) = 0 \end{cases}$$ (3.174)

and the initial conditions

$$\begin{cases} u(x,0) = f(x) \\ \dfrac{\partial u}{\partial t}(x,0) = g(x) \end{cases},$$ (3.175)

where $f(x)$ and $g(x)$ describe the displacement and velocity of the string at time $t=0$, respectively. Equations (3.172) to (3.174) characterise the string and its configuration. The properties of the system, ie, its natural frequencies and mode shapes of vibrations, are thus determined by (3.172) to (3.174) alone. To obtain the actual displacement $u(x,t)$ we must use the initial conditions (3.175) as well.

Let us now determine the natural frequencies and mode shapes of the string. As done in the previous sections, we assume that the string vibrates with harmonic motion of the form

$$u(x,t) = v(x)\sin\omega t.$$ (3.176)

Note that by this assumption we have separated the variables x and t. Substituting (3.176) in (3.172) gives

$$-\omega^2 v \sin\omega t = c^2 v'' \sin\omega t,$$ (3.177)

or

$$v'' + \left(\frac{\omega}{c}\right)^2 v = 0, \tag{3.178}$$

where primes denote derivatives with respect to x. Substituting (3.176) in (3.174) gives

$$\begin{cases} v(0)\sin\omega t = 0 \\ v(L)\sin\omega t = 0 \end{cases}. \tag{3.179}$$

An assumption that $\sin\omega t = 0$ leads through (3.176) to a trivial solution $u(x,t)=0$. Such a solution is not of interest since it predicts the dynamics of the string only in the case where the initial conditions (3.175) vanish. We thus conclude that (3.179) implies

$$\begin{cases} v(0) = 0 \\ v(L) = 0. \end{cases} \tag{3.180}$$

Combining (3.178) and (3.180) we obtain the eigenvalue problem

$$\begin{cases} v'' + \lambda v = 0 \\ v(0) = 0, \quad v(L) = 0, \end{cases} \tag{3.181}$$

where

$$\lambda = \left(\frac{\omega}{c}\right)^2. \tag{3.182}$$

In analogy with the discrete eigenvalue problem $(\mathbf{K} - \lambda\mathbf{M})\phi = \mathbf{o}$, the continuous eigenvalue problem (3.181) requires evaluation of the eigenvalues λ_i and their associated eigenfunctions $v_i(x) \neq 0$. The general solution of the differential equation in (3.181) is

$$v(x) = A_1 \sin\sqrt{\lambda}x + A_2 \cos\sqrt{\lambda}x, \tag{3.183}$$

where A_1 and A_2 are arbitrary constants. The left boundary condition in (3.181) yields

$$v(0) = A_2 = 0, \tag{3.184}$$

so that (3.183) reduces to

$$v(x) = A_1 \sin \sqrt{\lambda} x. \tag{3.185}$$

The right boundary condition in (3.181) gives

$$v(L) = A_1 \sin \sqrt{\lambda} L = 0. \tag{3.186}$$

Selecting $A_1 = 0$ leads to the trivial solution $v = u = 0$. We therefore choose

$$\sin \sqrt{\lambda} L = 0, \tag{3.187}$$

and obtain the eigenvalues λ_i from

$$\sqrt{\lambda_i} L = i\pi, \quad i = 1,2,3,\dots. \tag{3.188}$$

The natural frequencies $(\omega_n)_i$ of the string are determined by using (3.182) to give

$$(\omega_n)_i = \frac{ic\pi}{L}, \quad i = 1,2,3,\dots \tag{3.189}$$

By (3.185) their associated eigenfunctions v_i are

$$v_i(x) = \sin \frac{i\pi}{L} x, \quad i = 1,2,3,\dots . \tag{3.190}$$

In the discrete case the eigenvectors can be scaled arbitrarily. Similarly, the eigenfunctions of the distributed parameter system are determined up to a scalar constant A_i (see (3.186)). In (3.190) we have used (3.186) with $A_1 = 1$.

We know that resonance occurs when a multi-degree-of-freedom system is excited by a harmonic force with frequency that is equal to a natural frequency. Similarly, the string vibrates in resonance when a harmonic force $F_0 \sin \omega_n t$ is applied to the string, where F_0 is some constant and ω_n is one of the natural frequencies (3.189) of the string.

We now give a physical interpretation to the mode shapes (or the eigenfunctions). If the initial conditions are

$$\begin{cases} u(x,0) = v_i(x) \\ \dfrac{\partial u}{\partial t}(x,0) = 0 \end{cases},$$

(3.191)

where v_i is the i-th mode shape of the system, then each point of the string vibrates in a single harmonic frequency $(\omega_n)_i$. More precisely the string vibrates in this case with the mode shape

$$u(x,t) = v_i(x)\sin\frac{ic\pi}{L}t.$$

(3.192)

The axially vibrating rod. Consider the axially vibrating rod, of density $\rho(x)$, cross-sectional area $A(x)$, fixed at $x=0$ and free to oscillate at $x=L$, as shown in Figure 3.25(a). Denote by $u(x, t)$ the displacement of the small element, lying originally in the interval $[x, x+dx]$. The free body diagram for an instant where $0<u(x, t)<u(x+dx, t)$ is shown in Figure 3.25(b).

Noting that the mass of the element is ρAdx, its acceleration $\partial^2 u/\partial t^2$, we find that Newton's second law gives

$$-P + \left(P + \frac{\partial P}{\partial x}dx \right) = \rho Adx\frac{\partial^2 u}{\partial t^2},$$

(3.193)

or

$$\frac{\partial P}{\partial x} = \rho A\frac{\partial^2 u}{\partial t^2}.$$

(3.194)

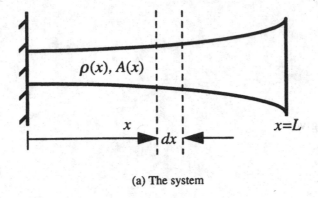

(a) The system

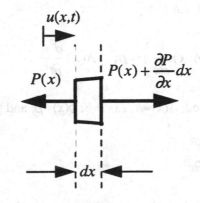

(b) Free body diagram

Figure 3.25   The axially vibrating rod

Recall the stress-strain relation

$$\sigma = E\varepsilon, \tag{3.195}$$

where σ is the applied stress, E is the modulus of elasticity and ε is the resulting strain. By definition

$$\sigma = \frac{P}{A} \tag{3.196}$$

and

$$\varepsilon = \frac{\partial u}{\partial x}.$$ (3.197)

Hence (3.195) gives

$$P = EA \frac{\partial u}{\partial x}.$$ (3.198)

Substituting (3.198) in (3.194) we obtain the differential equation of motion for the non-uniform rod

$$\frac{\partial}{\partial x}\left(E(x)A(x)\frac{\partial u}{\partial x} \right) = \rho(x)A(x)\frac{\partial^2 u}{\partial t^2}.$$ (3.199)

For a uniform rod, $A(x)=A$, $E(x)=E$, $\rho(x)=\rho$, and (3.199) simplifies to

$$\frac{\partial^2 u}{\partial t^2} = \frac{E}{\rho}\frac{\partial^2 u}{\partial x^2}.$$ (3.200)

We denote the speed of wave propagation in the uniform rod by

$$c = \sqrt{\frac{E}{\rho}}$$ (3.201)

and write (3.200) in the form

$$\frac{\partial^2 v}{\partial t^2} = \chi^2 \frac{\partial^2 v}{\partial \xi^2},$$ (3.202)

which is identical to the equation governing the motion of the string (3.172). Note however, that for the rod c is given by (3.201), whereas for the string c is given by (3.173).

    The boundary conditions for the fixed-free configuration of the rod imply that there is no displacement at $x=0$, and no axial force at $x=L$ (since there is no physical contact from the right at the free end). In view of (3.198)

$P(L)=0$ when the derivative of u with respect to x vanishes. The boundary conditions are, therefore,

$$\begin{cases} u(0,t) = 0 \\ \dfrac{\partial u(L,t)}{\partial x} = 0 \end{cases} . \tag{3.203}$$

To determine the eigenvalue problem associated with the differential equation (3.202) and the boundary conditions (3.203) we assume harmonic motion of the form

$$u(x,t) = v(x) \sin \omega t \tag{3.204}$$

and, upon substitution of (3.204) in (3.202) and (3.203), we obtain

$$\begin{cases} v''+\lambda v = 0 , \quad \lambda = \dfrac{\omega^2}{c^2} \\ v(0) = 0 \\ v'(L) = 0 \end{cases} . \tag{3.205}$$

The general solution (3.183) with the left boundary condition yields

$$v = A_1 \sin\sqrt{\lambda}x . \tag{3.206}$$

The right boundary condition gives

$$v'(L) = A_1 \sqrt{\lambda} \cos\sqrt{\lambda}L = 0 . \tag{3.207}$$

Since $A_1=0$ or $\lambda=0$ lead to the unwanted trivial solution $u(x,t)=0$, we determine the frequency equation

$$\cos\sqrt{\lambda}L = 0, \tag{3.208}$$

with the roots

$$\sqrt{\lambda_i}L = \frac{\pi}{2}, \frac{3\pi}{2}, \frac{5\pi}{2}, \dots \tag{3.209}$$

Using (3.205) we find the natural frequencies of the fixed-free uniform rod

$$(\omega_n)_i = \frac{(2i-1)c\pi}{2L}, \qquad i = 1,2,3,\dots \tag{3.210}$$

where c is given by (3.201). Choosing $A_1=1$ arbitrarily the eigenfunctions (3.206) are

$$v_i = \sin\frac{(2i-1)\pi x}{2L}, \qquad i = 1,2,3,\dots \tag{3.211}$$

We could use other values for A_1 and obtain other values for the eigenfunctions v_i.

Test Example 3.4
Determine the frequency equation and mode shapes of the uniform axially vibrating rod of cross sectional area A, modulus of elasticity E and density ρ, fixed at $x=0$ and spring supported at $x=L$, as shown in Figure 3.26(a).

Solution
The rod at static equilibrium, and in motion, is shown in Figure 3.26(b). This illustration shows that the right boundary condition is

$$\sigma(L,t) = -\frac{ku(L,t)}{A}, \tag{3.212}$$

where the minus sign in (3.212) is due to the fact that compressive stress is negative. By (3.195) and (3.197) we have

$$\sigma(L,t) = E\frac{\partial u(L,t)}{\partial x}, \tag{3.213}$$

so that the right boundary condition is

$$E\frac{\partial u(L,t)}{\partial x} = -\frac{ku(L,t)}{A}. \tag{3.214}$$

It follows that this rod is characterised by the differential equation

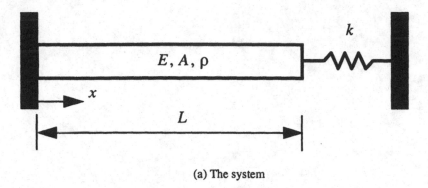

(a) The system

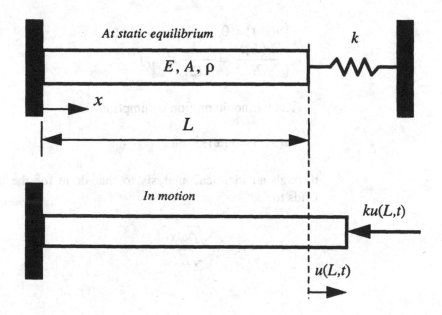

(b) The boundary condition

Figure 3.26 The spring supported uniform rod

$$\frac{\partial^2 u}{\partial t^2} = \frac{E}{\rho} \frac{\partial^2 u}{\partial x^2} ,$$

(3.215)

with the boundary conditions

$$\begin{cases} u(0,t) = 0 \\ \dfrac{\partial u(L,t)}{\partial x} = -\dfrac{k}{EA} u(L,t) \end{cases}.$$

The harmonic motion assumption

$$u(x,t) = v(x)\sin\omega t,$$

through an identical analysis to that done for the uniform
leads to

$$v(x) = A_1 \sin\frac{\sqrt{\rho}\,\omega x}{\sqrt{E}},$$

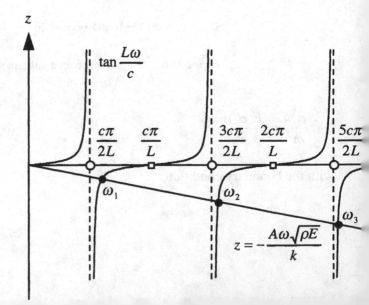

where $v(x)$ is a mode shape and ω is a natural frequency. Substituting (3.217) and (3.218) in the right boundary condition of (3.216) gives

$$\frac{\sqrt{\rho}\omega}{\sqrt{E}} A_1 \cos\frac{\sqrt{\rho}\omega L}{\sqrt{E}}\sin\omega t = -\frac{k}{EA} A_1 \sin\frac{\sqrt{\rho}\omega L}{\sqrt{E}}\sin\omega t, \qquad (3.219)$$

or

$$\tan\frac{\sqrt{\rho}L\omega}{\sqrt{E}} = -\frac{\sqrt{\rho E}\, A\omega}{k}. \qquad (3.220)$$

The mode shapes for the rod are

$$v(x) = \sin\frac{\sqrt{\rho}\omega x}{\sqrt{E}}, \qquad (3.221)$$

where ω is any root of the frequency equation (3.220).

Denote $z = -A\omega\sqrt{\rho E}\,/k$. Then the natural frequencies ω_i of the rod are the intersections of z and $\tan(L\omega/c)$, as shown in Figure 3.27. We see from this graph that there is, as expected, an infinite number of natural frequencies for the distributed parameter rod. Note also that the fixed-free rod is a special case where $k \to 0$. The fixed-fixed configuration is another extreme, where $k \to \infty$. It also follows from the figure that the effect of increasing the spring constant k is to increase all the natural frequencies of the system. It was observed by Lord Rayleigh that this phenomenon applies not only for the vibrating rod but to all linear vibratory systems, finite dimensional or distributed parameter systems.

3.8 Exercises

1. The two particles of mass $m_A=1$ kg and $m_B=3$ kg, shown in Figure 3.28, are initially at rest on a smooth horizontal plane. They are connected by an inextensible string, of length $L=2$ m, that passes through a small smooth ring that is fixed to the plane. Particle A is then projected at a right angle to the string with velocity $v_0=2$ m/s.
   (a) Write the differential equation of motion for particle A.
   (b) Determine the velocity of particles A and B as a function of the distance r_A of particle A from the ring.

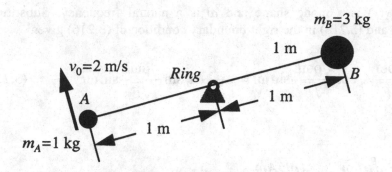

Figure 3.28 Two particles on a horizontal plane

2. Consider the three dimensional motion of the particle of mass m which is attached to a spring of free length L and spring constant k, as shown in Figure 3.29.
   (a) Write the three equations of motion for the particle.
   (b) Determine the linear system of equations governing the small oscillations of the particle about its stable equilibrium position.
   (c) Determine the three natural frequencies of the small oscillations.

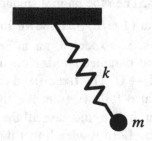

Figure 3.29 The flexible pendulum

3. A ring of radius R rotates with constant angular velocity $\dot\theta$ about a horizontal axis, as shown in Figure 3.30. A small bead P of mass m slides on the ring without friction.
   (a) Determine the equation of motion for the bead.
   (b) Determine the equilibrium positions for the bead and analyse their stability.
   (c) What is the frequency of oscillation for small motion of the bead about the stable equilibrium position?

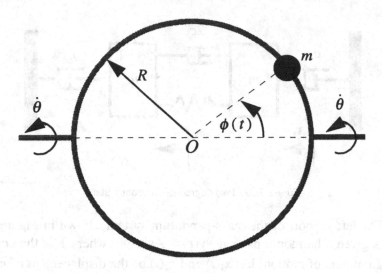

Figure 3.30 A bead on rotating ring

4. The initial conditions for the two degree-of-freedom system, shown in Figure 3.31, are

$$\begin{pmatrix} x_1(0) \\ x_2(0) \end{pmatrix} = \begin{pmatrix} 1 \\ 0 \end{pmatrix} \quad \text{and} \quad \begin{pmatrix} \dot{x}_1(0) \\ \dot{x}_2(0) \end{pmatrix} = \begin{pmatrix} 0.5 \\ 0 \end{pmatrix},$$

where $x_1(t)$ and $x_2(t)$ are the displacements of the masses from equilibrium.

(a) Write down the differential equations of motion for the system.

(b) What is the damped natural frequency of oscillation?

(c) Determine the relation between m, k and c for which the system does not vibrate (ie, the two modes of vibration are overdamped).

(d) Determine the response, $x_1(t)$ and $x_2(t)$, of the system when $k=4$, $m=2$ and $c=1$.

(e) With the same parameters as in part (d), what is the final positions of the masses?

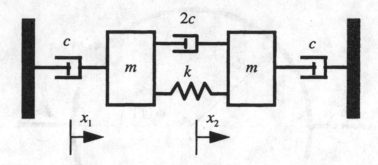

Figure 3.31 Two degree-of-freedom system

5. The left support of the mass-pendulum system, shown in Figure 3.32, is given a harmonic motion $y(t) = Y \sin \omega t$, where Y is the constant amplitude of motion. Let $x_1(t)$ and $x_2(t)$ be the displacement of m_1 and m_2 from equilibrium, respectively, and assume small vibrations about equilibrium.

(a) Write down the differential equations of motion for this system.

(b) Determine the frequency $\omega \neq 0$ for which the steady state response of the mass m_1 vanishes.

Assume now that the system is excited with the frequency ω obtained in part (b) above.

(c) What is the maximal force F_{MAX} transmitted to the left support at steady state?

(d) At steady state, for the instant t_1 when the displacement of the left support is $y(t_1)=Y$, what is the angle $\theta(t_1)$ of the pendulum?

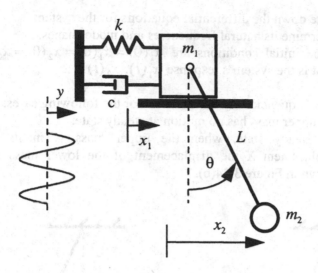

Figure 3.32 Mass-pendulum system

6. The system shown in Figure 3.33 consists of three point masses. At static equilibrium, the distance between two neighbouring masses is L. Two external opposing harmonic forces $f_L(t) = -f_R(t) = \cos\omega t$ are applied on the left and right masses as shown in the figure. The system is initially at rest.

   (*a*) Write down the differential equation of motion and determine its natural frequencies.

   (*b*) If $\omega = 0.5\sqrt{k/m}$, what is the minimal distance d between the left mass and the middle mass?

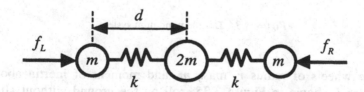

Figure 3.33 Three degree-of-freedom system

7. Consider the small vibrations about equilibrium of the double pendulum shown in Figure 3.34(a). The upper mass is excited by a harmonic force $f(t) = \cos\omega t$, as shown in the figure.

(*a*) Write down the differential equations for the system.

(*b*) Determine its natural frequencies and mode shapes.

(*c*) If the initial conditions are $x_1(0) = \dot{x}_1(0) = x_2(0) = \dot{x}_2(0) = 0$, what is the system's response $x_1(t)$, $x_2(t)$?

Find the frequencies of excitation ω for the following cases:

(*d*) The upper mass has no motion at steady state.

(*e*) At steady state, when the upper mass is in its maximal displacement X the displacement of the lower mass is $-X$, as shown in Figure 3.34(b).

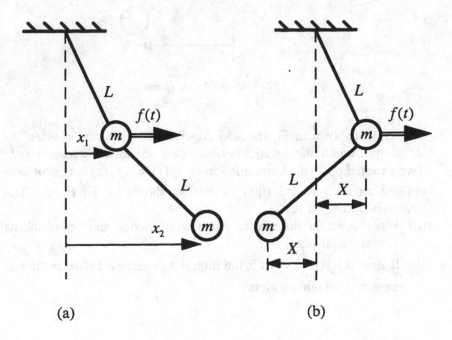

(a)   (b)

Figure 3.34 Double pendulum system

8. The wheels of radius r, mass m, and moment of inertia about the centre I, shown in Figure 3.35, roll on the ground without slipping. Their centres are supported by dampers of constant c; a spring of constant k connects the two centres of the wheels, as shown in the figure. The initial conditions are

$$\begin{pmatrix} x_1(0) \\ x_2(0) \end{pmatrix} = \begin{pmatrix} 0 \\ 0 \end{pmatrix}, \quad \begin{pmatrix} \dot{x}_1(0) \\ \dot{x}_2(0) \end{pmatrix} = \begin{pmatrix} 2 \\ 1 \end{pmatrix},$$

where $x_1(t)$ and $x_2(t)$ are the x-position of the centres of the wheels.

(a) Write down the differential equations of motion for the system.

Suppose that $r=1$ m, $m=2$ kg, $I=0.5$ kg·m2, $k=4$ N/m, $c=1$ N·sec/m.

(b) Determine the response $x_1(t)$, $x_2(t)$, of the system.
(c) What are the final positions of the wheels?

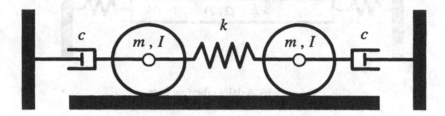

Figure 3.35 Wheel-spring-damper system

9. Consider the axially vibrating uniform rod of length L, cross sectional area A, density ρ and modulus of elasticity E. The rod is free to vibrate at the left end and spring supported at its right end, as shown in Figure 3.36(a).

   (a) Write down the differential equation of motion and the boundary conditions for the rod.
   (b) Determine the frequency equation of the rod.
   (c) Let ω_i be the i-th lowest natural frequency of the rod. Approximate ω_n for large integer n. -

A spring of constant k is added to the left end of the rod, as shown in Figure 3.37(b). It is found that the lowest natural frequency of the rod in its new configuration is $\pi/4$.

(d) Determine the spring constant k, assuming that $A=\rho=E=L=1$.

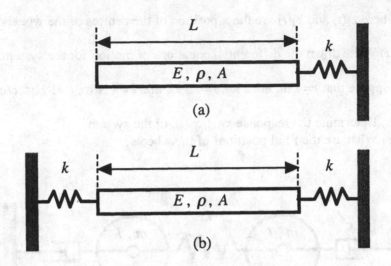

(a)

(b)

Figure 3.36 Axially vibrating rods

10. Consider the axially vibrating uniform rod of length L, cross
    sectional area A, density ρ and modulus of elasticity E. The rod is
    supported by a spring of constant k at the left end, and a mass M is
    attached to its right end, as shown in Figure 3.37.
    (a) Write down the differential equation of motion and the boundary
        conditions for the rod.
    (b) Determine the frequency equation for the rod.
    (c) Let ω_i be the i-th lowest natural frequency of the rod.
        Approximate ω_n for large integer n.

    Assume that $A=\rho=E=L=1$ and $M=3$. If the lowest natural frequency
    of the rod is $\pi/10$,
    (d) what is the spring constant k?

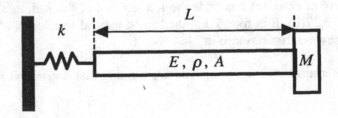

Figure 3.37 Rod-mass-spring system

Bibliography

1. Inman, D.J., *Engineering Vibration*, Prentice Hall, New Jersey, 1994.
2. Meirovitch, L, *Elements of Vibration Analysis*, McGraw-Hill, New York, Second Edition, 1986.
3. Meriam, J. L., and Kraige, L.G., *Engineering Mechanics: Dynamics*, John Wiley & Sons, New York, Fourth Edition, 1997.

Chapter 4

DISCRETE MODELS.

4.1. Introduction

The models described in Chapter 3 are continuous time models. The time variable is a continuous function and the models are expressed in terms of differential equations of the form

$$L[u(t)] = f(t)$$

where $u(t)$ is the response of the system to the excitation $f(t)$, and $L[u(t)]$ is some differential operator acting on $u(t)$. In some applications, for example digital system control, the dynamics of the system is sampled with sampling rate Δt. The response of the system is presented in terms of a discrete function, ie. a finite vector $\mathbf{u} = (u_1, u_2, ..., u_m)^T$ where $u_k = u(k\Delta t)$, $k = 1, 2, ..., m$. The input may be also discrete $\mathbf{f} = (f_1, f_2, ..., f_m)^T$ where $f_k = f(k\Delta t)$ is the input force applied at $t = k\Delta t$. If there exists a linear relation between the exciting force $\mathbf{f}$ and the response $\mathbf{u}$ then the input-output relation can be expressed in terms of a linear difference equation of order n

$$\sum_{j=0}^{n} a_j(k)u_{k+j} = f_k, \quad k = 1, 2, ..., m - j,$$

where $a_j(k)$ are the coefficients of the difference equation. The finite difference method allows us to replace the differential equations associated with the continuous time model by difference equations representing the dynamics of the discrete time system. An introduction to finite difference approximation is presented in Section 4.2. The solution to initial value linear difference equations may be simulated easily by the computer. In some cases we may obtain an analytical closed form solution of the discrete model. A method for determining analytical solutions to linear difference equations and systems of difference equations with constant coefficients is presented in Section 4.3.

Continuous time systems may be stable, unstable or marginally stable. In Section 3.4 the stability of continuous time systems has been characterised in terms of the location of the poles in the s-plane. A similar approach will be developed in Section 4.4 where the stability of the system will be analysed by considering the location of its poles in the z-plane. It will then be demonstrated in Section 4.5 that, depending on the sampling rate, a stable continuous time system my be presented by an unstable discrete time system. A relation between the sampling rate and the poles of the continuous system which ensures the stability of the discrete time model associated with the stable continuous time system is then developed.

Another important application of finite difference and discrete models is in predicting the eigenvalues and eigenfunctions of systems governed by partial differential equations. In this case the discretisation is done on the spatial variable, rather than on the time. To demonstrate this application we derive a finite difference model for an axially vibrating rod in Section 4.6 and approximate some of its natural frequencies and mode shapes.

4.2. Finite Difference Approximation

Consider a continuous smooth function $u(x)$ such as that shown in Figure 4.1. Let x_{k-1}, x_k, x_{k+1} be three points on the x-axis, distance h apart. Denote $u_j = u(x_j)$. We wish to approximate the derivative $u' = du/dx$ at $x = x_k$ in terms of the function values u_{k-1}, u_k and u_{k+1}. Since $u'(x_k)$ is the slope of the tangent S-T shown in the figure, we may use the following three possible approximations:

backward scheme

$$u'_k \cong \frac{u_k - u_{k-1}}{h},$$
(4.1)

forward scheme

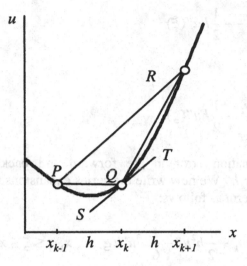

Figure 4.1. The first derivative

$$u'_k \cong \frac{u_{k+1} - u_k}{h},$$ (4.2)

and *central* scheme

$$u'_k \cong \frac{u_{k+1} - u_{k-1}}{2h},$$ (4.3)

with the graphical interpretation of approximating the tangent *S-T* by the chords *P-Q*, *Q-R* or *P-R*, shown in Figure 4.1, respectively.

From inspection of the figure it appears that the central scheme provides the better approximation for the function shown. Indeed, the series expansions of u_{k-1} and u_{k+1} about x_k

$$u_{k-1} = u_k - hu'_k + \frac{1}{2}h^2 u''(\xi), \quad x_{k-1} \le \xi \le x_k,$$ (4.4)

$$u_{k+1} = u_k + hu'_k + \frac{1}{2}h^2 u''(\varsigma), \quad x_k \le \varsigma \le x_{k-1},$$ (4.5)

give

$$u'_k = \frac{u_k - u_{k-1}}{h} + \frac{1}{2}hu''(\xi) \qquad (4.6)$$

and

$$u'_k = \frac{u_{k+1} - u_k}{h} - \frac{1}{2}hu''(\varsigma). \qquad (4.7)$$

The approximation error in the forward and backward scheme is therefore of order h. We now write the series expansions using three terms and a remainder term as follows:

$$u_{k-1} = u_k - hu'_k + \frac{1}{2}h^2u''_k - \frac{1}{6}h^3u'''(\xi), \quad x_{k-1} \le \xi \le x_k, \qquad (4.8)$$

$$u_{k+1} = u_k + hu'_k + \frac{1}{2}h^2u''_k + \frac{1}{6}h^3u'''(\varsigma), \quad x_k \le \varsigma \le x_{k+1}. \qquad (4.9)$$

On subtracting the first equation from the second and dividing the result by $2h$, we obtain

$$\frac{u_{k+1} - u_{k-1}}{2h} = u'_k + \frac{1}{12}h^2\left(u'''(\xi) + u'''(\varsigma)\right). \qquad (4.10)$$

The error in replacing the derivative by a central difference scheme is thus of order h^2.

We now develop a difference scheme for approximating the second derivative u''_k in terms of u_{k-1}, u_k and u_{k+1}. Let x_A and x_B be the midpoints of the intervals $[x_{k-1}, x_k]$ and $[x_k, x_{k+1}]$, respectively, as shown in Figure 4.2. Denote $u_A = u(x_A)$ and $u_B = u(x_B)$. Then using the central difference scheme we have

$$u'_A \cong \frac{u_k - u_{k-1}}{h} \qquad (4.11)$$

and

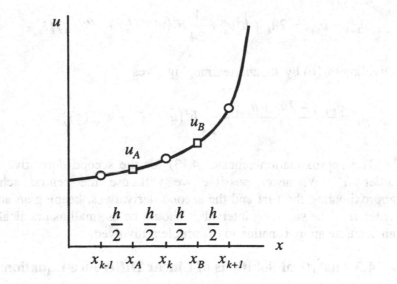

Figure 4.2. The second derivative

$$u'_B \cong \frac{u_{k+1} - u_k}{h}.$$ (4.12)

The second derivative may thus be expressed in the form

$$u''_k = \frac{u'_B - u'_A}{h} = \frac{u_{k-1} - 2u_k + u_{k+1}}{h^2}.$$ (4.13)

To analyse the error associated with this scheme we expand u_{k-1} and u_{k+1} about x_k

$$u_{k-1} = u_k - hu'_k + \frac{1}{2}h^2 u''_k - \frac{1}{6}h^3 u'''_k + \frac{1}{24}h^4 u''''(\xi),$$ (4.14)

$$x_{k-1} \le \xi \le x_k,$$

$$u_{k+1} = u_k + hu'_k + \frac{1}{2}h^2 u''_k + \frac{1}{6}h^3 u'''_k + \frac{1}{24}h^4 u''''(\varsigma),$$ (4.15)

$$x_k \le \varsigma \le x_{k+1}.$$

Adding the two expressions in (4.14) and (4.15) we have

$$u_{k-1} + u_{k+1} = 2u_k + h^2 u_k'' + \frac{1}{24} h^4 \left(u''''(\xi) + u''''(\varsigma) \right). \tag{4.16}$$

Dividing (4.16) by h^2 and rearranging gives

$$u_k'' = \frac{u_{k-1} - 2u_k + u_{k+1}}{h^2} - \frac{1}{24} h^2 \left(u''''(\xi) + u''''(\varsigma) \right). \tag{4.17}$$

The approximation scheme (4.13) for the second derivative is thus of order h^2. Whenever possible we will use the central schemes for approximating the first and the second derivatives, keeping an accuracy of order h^2. The sampling interval h should be as small as practical to obtain an accurate approximation to the problem involved.

4.3 Analytical Solutions of Linear Difference Equations

A single difference equation. Consider the homogeneous difference equation

$$\sum_{j=0}^{n} a_j u_{k+j} = 0, \quad k = 1,2,... \tag{4.18}$$

with constant coefficients a_j, $j = 0,1,...,n$. We try a solution of the form

$$u_k = z^k \tag{4.19}$$

and obtain, upon substituting in (4.18)

$$\sum_{j=0}^{n} a_j z^{k+j} = z^k \sum_{j=0}^{n} a_j z^j = 0. \tag{4.20}$$

Denote the characteristic polynomial of degree n by

$$P_n(z) = \sum_{j=0}^{n} a_j z^j . \tag{4.21}$$

Then $P_n(z)$ has generally n roots $z_1, z_2, ..., z_n$, and using (4.19) a non-trivial solution of (4.18) is given by

$$u_k = \sum_{j=1}^{n} A_j z_j^k \qquad (4.22)$$

where A_j, $j = 1, 2, \ldots, n$ are arbitrary constants. If the roots of $P_n(z)$ are distinct, ie. $z_i \neq z_j$ when $i \neq j$, then (4.22) is said to be the general solution of (4.18). In this case the coefficients A_j are determined uniquely by n initial conditions u_j, $j = 1, 2, \ldots, n$ via the solution of the Vandermonde system of equations

$$\begin{bmatrix} z_1 & z_2 & \cdots & z_n \\ z_1^2 & z_2^2 & \cdots & z_n^2 \\ \vdots & \vdots & & \vdots \\ z_1^n & z_2^n & \cdots & z_n^n \end{bmatrix} \begin{pmatrix} A_1 \\ A_2 \\ \vdots \\ A_n \end{pmatrix} = \begin{pmatrix} u_1 \\ u_2 \\ \vdots \\ u_n \end{pmatrix}. \qquad (4.23)$$

In a similar way we may determine A_j by any other set of n conditions u_j which are not necessarily initial to the dynamics of the system.

Test Example 4.1

Determine the solution of the second order difference equation

$$2u_k - 3u_{k+1} + u_{k+2} = 0, \qquad k = 1, 2, \ldots, 10, \qquad (4.24)$$

with the initial and terminal conditions

$$u_1 = 0, \quad u_{10} = -1. \qquad (4.25)$$

Solution.

The characteristic polynomial for this problem is

$$P_2(z) = z^2 - 3z + 2 = (z - 1)(z - 2).$$

Hence with $z_1 = 1$ and $z_2 = 2$ the general solution of (4.24) is

$$u_k = A_1 + A_2 2^k. \qquad (4.26)$$

The Vandermonde system which determines the solution here is

$$\begin{bmatrix} 1 & 2 \\ 1 & 2^{10} \end{bmatrix} \begin{pmatrix} A_1 \\ A_2 \end{pmatrix} = \begin{pmatrix} 0 \\ -1 \end{pmatrix}$$

which gives

$$A_1 = (2^9 - 1)^{-1} \text{ and } A_2 = (2 - 2^{10})^{-1}. \qquad\blacksquare$$

If for some integer p, $0 \le p \le n$, z_p is a root of $P_n(z)$ with multiplicity $q > 1$ then the solution (4.22) does not contain n linearly independent functions. Consequently the system (4.23) is singular and n conditions do not determine the coefficients A_j, $j = 1,2,...,n$ uniquely. It is easy to confirm by direct substitution in (4.18), however, that

$$u_k = k^r z_p^k \qquad\qquad\qquad (4.27)$$

is another solution of the problem for each $r = 1,2,..,q$. We may thus obtain using (4.19) and (4.27) n linearly independent functions which represent the general solution of (4.27). The following example demonstrates such a case.

Test Example 4.2
Find the solution of

$$-12u_k + 16u_{k+1} - 7u_{k+2} + u_{k+3} = 0, \qquad\qquad (4.28)$$

with the initial conditions $u_0 = 0$, $u_1 = 0$, $u_2 = 1$.

Solution
The characteristic polynomial associated with (4.28) is

$$P_3(z) = z^3 - 7z^2 + 16z - 12$$
$$= (z_1 - 2)^2 (z_2 - 3).$$

The general solution of (4.28) is thus

$$u_k = (A_1 k + A_2)2^k + A_3 3^k.$$

The initial conditions lead to the linear system of equations

$$\begin{bmatrix} 0 & 1 & 1 \\ 2 & 2 & 3 \\ 8 & 4 & 9 \end{bmatrix} \begin{pmatrix} A_1 \\ A_2 \\ A_3 \end{pmatrix} = \begin{pmatrix} 0 \\ 0 \\ 1 \end{pmatrix},$$

which has the solution $A_1 = -0.5$, $A_2 = -1$, $A_3 = 1$. ∎

The solution to the non-homogeneous linear difference equation

$$\sum_{j=0}^{n} a_j u_{k+j} = f_k, \quad k = 1,2,\dots \tag{4.29}$$

where f_k is some function of k may be determined by superposition of the general solution of the homogeneous problem and a particular solution of (4.29). The process is demonstrated in the following example.

Test Example 4.3
Determine the general solution of the second order difference equation

$$2u_k - 3u_{k+1} + u_{k+2} = -2k + 1. \tag{4.30}$$

Solution.
The general solution of the associated homogeneous problem has been found in Test Example 4.1, namely, equation (4.26). We need to find a particular solution v_k to

$$2v_k - 3v_{k+1} + v_{k+2} = -2k + 1. \tag{4.31}$$

We try a particular solution of the form

$$v_k = C_2 k^2 + C_1 k + C_0,$$

and substitute it in (4.31) to give

$$2(C_2 k^2 + C_1 k + C_0)$$
$$-3(C_2(k+1)^2 + C_1(k+1) + C_0)$$
$$+C_2(k+2)^2 + C_1(k+2) + C_0 = -2k + 1,$$

which upon simplification gives

$$-2C_2 k + C_2 - C_1 = -2k + 1.$$

This equation is satisfied for all values of k if and only if

$$\begin{cases} -2C_2 = -2 \\ C_2 - C_1 = 1. \end{cases}$$

We thus have $C_2 = 1$, $C_1 = 0$ and C_0 may be chosen arbitrarily. A particular solution to (4.31) may therefore be chosen as

$$v_k = k^2$$

and, in view of (4.26), the general solution of (4.30) is given by

$$u_k = A_1 + A_2 2^k + k^2.$$ ∎

A system of difference equations.

A linear system of simultaneous difference equations with constant coefficients of arbitrary order can always be reduced to an equivalent system of first order difference equations of the form

$$\mathbf{A}\mathbf{u}[k] - \mathbf{B}\mathbf{u}[k+1] = \mathbf{o}, \tag{4.32}$$

where $\mathbf{A}$ and $\mathbf{B}$ are $n \times n$ matrices and $\mathbf{u}[k]$ is an n-vector defining the solution at the k-th step. We try a solution of the form

$$\mathbf{u}[k] = z^k \mathbf{v}, \tag{4.33}$$

where $\mathbf{v}$ is a constant vector, and obtain

$$z^k (\mathbf{A} - z\mathbf{B})\mathbf{v} = \mathbf{o}. \tag{4.34}$$

Hence, non-trivial solutions of (4.32) may be obtained by solving the generalised eigenvalue problem

$$\mathbf{A}\mathbf{v} = \lambda \mathbf{B}\mathbf{v}. \tag{4.35}$$

If A and B are invertible and the system (4.35) is non-defective then n distinct eigenpairs $\{\lambda_i, v_i\}$ may be obtained. The poles of the system are $z_i = \lambda_i$, $i = 1, 2, \ldots, n$. The eigenvector v_i is called the i-th mode of the system. The expression

$$u[k] = c z_i^k v_i \qquad (4.36)$$

satisfies (4.32) for an arbitrary constant c. By the linearity of the problem, the general solution of (4.32) is given in terms of the linear combination

$$u[k] = \sum_{i=1}^{n} c_i z_i^k v_i . \qquad (4.37)$$

The n arbitrary constants c_i are obtained by n initial conditions that determine a unique solution to the problem when the z_i are distinct.

The following example demonstrates the method described above.

Test Example 4.4
Solve the system of two simultaneous equations

$$\begin{cases} x_k + 2y_{k+1} + 3x_{k+3} = 0 \\ y_k + 4y_{k+1} + 6x_{k+2} = 0. \end{cases} \qquad (4.38)$$

Solution
The system is first realised by a first order matrix system. Noting that the order of x is three and y is of first order we define the vector of unknowns

$$u[k] = \begin{pmatrix} x_k & x_{k+1} & x_{k+2} & y_k \end{pmatrix}^T . \qquad (4.39)$$

Hence

$$u[k+1] = \begin{pmatrix} x_{k+1} & x_{k+2} & x_{k+3} & y_{k+1} \end{pmatrix}^T \qquad (4.40)$$

and we may write for example

$$
\begin{bmatrix} 0 & 1 & 0 & 0 \\ 0 & 0 & 1 & 0 \\ 1 & 0 & 0 & 0 \\ 0 & 0 & 6 & 1 \end{bmatrix} \begin{pmatrix} x_k \\ x_{k+1} \\ x_{k+2} \\ y_k \end{pmatrix} - \begin{bmatrix} 1 & 0 & 0 & 0 \\ 0 & 1 & 0 & 0 \\ 0 & 0 & -3 & -2 \\ 0 & 0 & 0 & -4 \end{bmatrix} \begin{pmatrix} x_{k+1} \\ x_{k+2} \\ x_{k+3} \\ y_{k+1} \end{pmatrix} = \begin{pmatrix} 0 \\ 0 \\ 0 \\ 0 \end{pmatrix}.
\tag{4.41}
$$

The first two equations are simply the identities $x_{k+1} - x_{k+1} = 0$ and $x_{k+2} - x_{k+2} = 0$. The last two equations are the same as in (4.38). We thus have the first order realisation

$$
\mathbf{A}u[k] - \mathbf{B}u[k+1] = \mathbf{o}
\tag{4.42}
$$

where

$$
\mathbf{A} = \begin{bmatrix} 0 & 1 & 0 & 0 \\ 0 & 0 & 1 & 0 \\ 1 & 0 & 0 & 0 \\ 0 & 0 & 6 & 1 \end{bmatrix}, \quad \mathbf{B} = \begin{bmatrix} 1 & 0 & 0 & 0 \\ 0 & 1 & 0 & 0 \\ 0 & 0 & -3 & -2 \\ 0 & 0 & 0 & -4 \end{bmatrix}.
\tag{4.43}
$$

The poles of the system are the eigenvalues of $\mathbf{A} - \lambda\mathbf{B}$

$$
z_1 = \bar{z}_2 = 0.6912 + i0.5362, \ z_3 = \bar{z}_3 = -0.3162 + i0.0945,
$$

and the modes are the eigenvectors

$$
\mathbf{v}_1 = \begin{pmatrix} 0.1822 + i0.5047 \\ -0.1447 + i0.4465 \\ -0.3394 + i0.2311 \\ 0.2500 - i0.5107 \end{pmatrix}, \quad \mathbf{v}_2 = \begin{pmatrix} 0.5047 + i0.1821 \\ 0.4465 - i0.1448 \\ 0.2309 - i0.3395 \\ -0.5106 + i0.2501 \end{pmatrix},
$$

$$
\mathbf{v}_3 = \begin{pmatrix} 0.5486 - i0.1381 \\ -0.1604 + i0.0955 \\ 0.0417 - i0.0454 \\ 0.7938 + i0.1057 \end{pmatrix}, \quad \mathbf{v}_4 = \begin{pmatrix} -0.1383 + i0.5486 \\ 0.0956 - i0.1604 \\ -0.0454 + i0.0417 \\ 0.1054 + i0.7938 \end{pmatrix}.
$$

The general solution to the problem is thus given by (4.37), which can be written equivalently in the following matrix form

$$\mathbf{u}[k] = \mathbf{V}\mathbf{Z}[k]\mathbf{c}$$

with

$$\mathbf{V} = [\mathbf{v}_1|\mathbf{v}_2|\mathbf{v}_3|\mathbf{v}_4],$$

$$\mathbf{Z}[k] = diag\{z_1^k, z_2^k, z_3^k, z_4^k\},$$

and the vector of arbitrary coefficients $\mathbf{c} = (c_1, c_2, c_3, c_4)^T$. If the state $\mathbf{u}[p]$ for a particular index p is given, then the coefficients c_i may be determined by

$$\mathbf{c} = (\mathbf{V}\mathbf{Z}[p])^{-1}\mathbf{u}[p]. \qquad\blacksquare$$

4.4 Stability

We have just shown that the solution to the linear homogeneous system of equations

$$\mathbf{A}\mathbf{u}[k] - \mathbf{B}\mathbf{u}[k+1] = \mathbf{o} \qquad\qquad (4.44)$$

is given by

$$\mathbf{u}[k] = \sum_{i=1}^{n} c_i z_i^k \mathbf{v}_i \qquad\qquad (4.45)$$

where z_i are the poles of the system, $\mathbf{v}_i$ are its modes, and c_i are arbitrary constants that may be determined by the initial conditions. We say that the system is stable if its response to arbitrary initial conditions vanishes for large k, ie. $\mathbf{u}[k] \to \mathbf{o}$ as $k \to \infty$. If the response of the system to arbitrary initial conditions increases without bound as k increases the system is said to be unstable. A system for which there exists a non-vanishing bounded steady state response is said to be marginally stable. These same definitions are applied to both discrete and continuous systems.

We note that if $\mathbf{A}$, $\mathbf{B}$ and the initial conditions are all real, the response of the system $\mathbf{u}[k]$ is real as well. Moreover, the eigenvalues of (4.35) which are the poles of the system are a self conjugate set in the sense that if z is a complex pole of the system then $\bar{z}$, its complex conjugate, is another pole of the system. Let z_p be a complex pole of (4.44). Then the mode associated with z_p is given by

$$c_p z_p^k \mathbf{v}_p,$$ (4.46)

and may be written equivalently as

$$c_p (a+ib)^k \mathbf{v}_p,$$ (4.47)

where a and b are real numbers. We may write the last expression in trigonometric form as

$$c_p r^k (\cos k\theta + i \sin k\theta) \mathbf{v}_p$$ (4.48)

with

$$r = \sqrt{a^2 + b^2}$$ (4.49)

$$\theta = \arctan \frac{b}{a}.$$ (4.50)

It is thus clear that the dynamics associated with this mode increases without bound if $r > 1$, approaches zero for $r < 1$ and oscillates with finite amplitude when $r = 1$. Since $r = |z_p|$ we may characterise the stability associated with $\mathbf{v}_p$ by using the absolute values of the pole instead of the magnitude of r. The system is stable if all its poles are stable, unstable if at least one of its poles is unstable, and marginally stable if some of its simple poles are marginally stable and all other poles are stable. We thus conclude that a system with n poles is stable if $|z_i| < 1$ for $i = 1,2,...,n$, and unstable if $|z_j| > 1$ for some j. For marginally stable systems $|z_i| \leq 1$ where the equality holds for some simple poles.

Following conventions we draw a unit circle about the origin of the complex plane and indicate the position of the poles in this plane, called the z-plane. The stability of the system is then determined by the position of the poles with respect to the stable region, the unit circle in the z-plane, as illustrated in Figure 4.3.

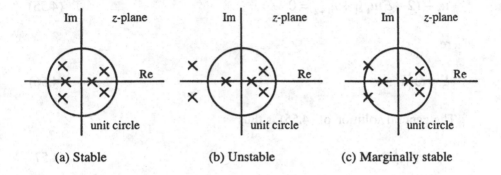

(a) Stable   (b) Unstable   (c) Marginally stable

Figure 4.3 Stability and location of poles in the z-plane

4.5. Discrete Time Models of Dynamic Systems

The simple oscillator.

The differential equation governing the free vibrations of the simple oscillator (see equation (3.18) with $c = 0$ and $k = \kappa$) is

$$m\ddot{u} + \kappa u = 0. \tag{4.51}$$

The analytical solution to this problem is

$$u(t) = A_1 \sin\omega_n t + A_2 \cos\omega_n t, \quad \omega_n = \sqrt{\frac{\kappa}{m}}, \tag{4.52}$$

where A_1 and A_2 are arbitrary constants which are determined by the initial conditions. We sample the time $t = kh$, where h is a constant time increment. Then by using the central difference scheme for the second derivative (4.13)

$$\ddot{u}(kh) = \frac{u_{k-1} - 2u_k + u_{k+1}}{h^2} \tag{4.53}$$

we arrive at the following discrete time model

$$\frac{m}{h^2}(u_{k-1} - 2u_k + u_{k+1}) + \kappa u_k = 0. \tag{4.54}$$

Equation (4.54) may be written equivalently as

$$u_k - (2 - \xi)u_{k+1} + u_{k+2} = 0 \tag{4.55}$$

where

$$\xi = \omega_n^2 h^2. \tag{4.56}$$

The general solution of (4.55) is thus

$$u_k = A_1 z_1^k + A_2 z_2^k \tag{4.57}$$

where z_1, z_2 are the roots of

$$P_2(z) = z^2 - (2 - \xi)z + 1 = 0, \tag{4.58}$$

given explicitly by

$$z_{1,2} = \frac{2 - \xi \pm \sqrt{\xi^2 - 4\xi}}{2}. \tag{4.59}$$

Since by its definition $\xi \geq 0$, the poles of the system are complex when $0 < \xi < 4$ and real otherwise. For the case where $0 < \xi < 4$ we have

$$|z_1| = |z_2| = \sqrt{z_1 z_2} = 1. \tag{4.60}$$

Note that $0 < \xi < 4$ implies that $h < 2/\omega_n$ by virtue of (4.56). We therefore find that the difference equation governing the motion is marginally stable if the time sampling interval is smaller than the so called *Nyquist sampling interval* $2\omega_n^{-1}$. When $\xi > 4$ the pole $z_2 < -1$ and the difference equation (4.55) is unstable. The location of the roots of the characteristic equation (4.58) is shown in Figures 4.4(a) and 4.4(b). The absolute values of the poles as functions of ξ are illustrated in Figure 4.5. The continuous time system is marginally stable since its poles are located on the imaginary axis of the s-plane. Its discrete counterpart (4.54) is marginally stable only if h is smaller than the Nyquist sampling interval. It is thus clear that if $h > 2\omega_n^{-1}$ the solution of the discrete model does not approximate the dynamics of the continuous system adequately. The dynamics of the continuous system versus the discrete models for two different sampling times h are shown in Figures 4.6(a) and 4.6(b). In Figure 4.6(a) $h = 1.5\omega_n^{-1}$ is used. The unstable case where $h = 2.05\omega_n^{-1}$ is shown in Figure 4.6(b).

Let us now investigate whether the solution of the discrete model is consistent with that of the continuous model when $h \to 0$. Suppose $0 \le h \le 2\omega_n^{-1}$. Then invoking (4.48)-(4.50) we obtain

$$u_k = B_1 \sin k\theta + B_2 \cos k\theta, \quad 0 \le h \le 2\omega_n^{-1}, \tag{4.61}$$

where B_1 and B_2 are arbitrary constants and

$$\theta = \arctan \frac{\sqrt{4\xi - \xi^2}}{2 - \xi}. \tag{4.62}$$

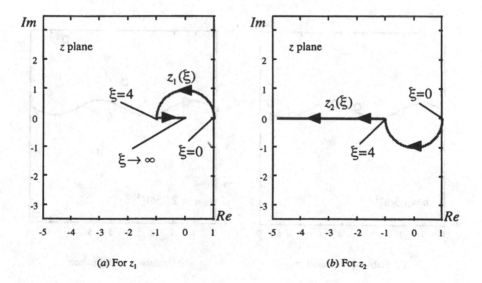

(a) For z_1 (b) For z_2

Figure 4.4. Root loci

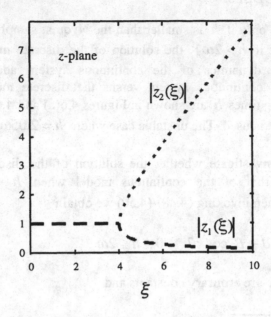

Figure 4.5. Magnitudes of z_1, z_2

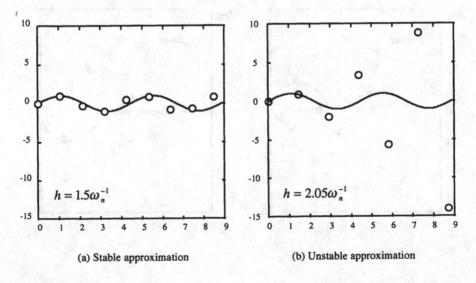

(a) Stable approximation   (b) Unstable approximation

Figure 4.6. The motion of a simple oscillator and its discrete time models approximations

In the limit where $\xi \to 0$ we have

$$\lim_{\xi \to 0} \arctan \frac{\sqrt{4\xi - \xi^2}}{2 - \xi} = \sqrt{\xi} \,. \tag{4.63}$$

But by (4.56)

$$\sqrt{\xi} = \omega_n h \tag{4.64}$$

and we obtain from (4.61)

$$u_k = B_1 \sin \omega_n kh + B_2 \cos \omega_n kh \,, \quad h \to 0 \,. \tag{4.65}$$

It follows from (4.52) that the solution to the discrete model is consistent with that of the continuous one when $h \to 0$.

Multi-degree of freedom systems.

Consider an n-degree-of-freedom system which is governed by the matrix differential equation

$$\mathbf{M\ddot{u}} + \mathbf{Ku} = 0 \tag{4.66}$$

such as that described in Section 3.5. Using (4.53) we obtain

$$h^{-2}\mathbf{M}\big(\mathbf{u}[k-1] - 2\mathbf{u}[k] + \mathbf{u}[k+1]\big) + \mathbf{Ku}[k] = 0 \tag{4.67}$$

where $\mathbf{u}[k]$ denotes the vector displacement of the system at the time kh.
We try a solution of the form

$$\mathbf{u}[k] = z^k \mathbf{v} \,, \tag{4.68}$$

where $\mathbf{v}$ is a constant vector, and obtain upon substitution in (4.67)

$$h^{-2} z^{k-1}(1 - 2z + z^2)\mathbf{Mv} + z^k \mathbf{Kv} = 0 \,. \tag{4.69}$$

Denote

$$\lambda = -\frac{h^{-2} z^{k-1}(1 - 2z + z^2)}{z^k} = -\frac{1 - 2z + z^2}{z h^2} \,. \tag{4.70}$$

Then (4.69) can be written in the form

$$(\mathbf{K} - \lambda \mathbf{M})\mathbf{v} = \mathbf{0}. \tag{4.71}$$

When $\mathbf{M}$ and $\mathbf{K}$ are positive definite symmetric matrices the generalised eigenvalue problem (4.71) has n real positive eigenvalues λ_i with corresponding eigenvectors $\mathbf{v}_i$, $i = 1, 2, \ldots, n$. It follows from (4.70) that

$$1 - (2 - \lambda h^2)z + z^2 = 0. \tag{4.72}$$

Hence each eigenvalue λ_i determines two poles z_{i1} and z_{i2}

$$z_{i1}, z_{i2} = \frac{2 - \xi_i \pm \sqrt{\xi_i^2 - 4\xi_i}}{2} \tag{4.73}$$

where

$$\xi_i = \lambda_i h^2. \tag{4.74}$$

It thus follows that for arbitrary constants A_1 and A_2 the i-th mode

$$(A_1 z_{i1}^k + A_2 z_{i2}^k)\mathbf{v}_i \tag{4.75}$$

is a solution of (4.67). Hence, by using superposition we obtain the general solution of (4.67)

$$\mathbf{u}[k] = \sum_{i=1}^{n} (A_i z_{i1}^k + B_i z_{i2}^k)\mathbf{v}_i. \tag{4.76}$$

The system is unstable if for one or more of its poles $\left| z_{i1,2} \right| > 1$. From the results of Section 4.5 we find that the i-th mode of the system is marginally stable when $h\sqrt{\lambda_i} < 2$ and unstable otherwise. Let λ_n be the largest eigenvalue of (4.71). Then the system is marginally stable if

$$h < \frac{2}{\sqrt{\lambda_n}} \tag{4.77}$$

and unstable if (4.77) is not satisfied. When the system is marginally stable its response may be expressed alternatively in the form

$$\mathbf{u}[k] = \sum_{i=1}^{n} (A_i \sin k\theta_i + B_i \cos k\theta_i) \mathbf{v}_i, \quad h\sqrt{\lambda_n} < 2, \tag{4.78}$$

where

$$\theta_i = \arctan \frac{\sqrt{4\xi_i - \xi_i^2}}{2 - \xi_i}. \tag{4.79}$$

Test Example 4.5
By using a discrete time model with constant time interval $h = 0.1$, determine the solution of the two-degree-of-freedom mass-spring system shown in Figure 4.7.

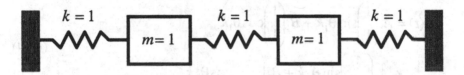

Figure 4.7. A two degree-of-freedom system

Solution
The system is modelled by the differential equations

$$\mathbf{M\ddot{u} + Ku = 0} \tag{4.80}$$

where

$$\mathbf{M} = \begin{bmatrix} 1 & 0 \\ 0 & 1 \end{bmatrix} \text{ and } \mathbf{K} = \begin{bmatrix} 2 & -1 \\ -1 & 2 \end{bmatrix}. \tag{4.81}$$

The eigenvalue problem $\mathbf{Kv} = \lambda \mathbf{Mv}$ has two eigenpairs

$$\left\{ 1, \begin{pmatrix} 1 \\ 1 \end{pmatrix} \right\} \text{ and } \left\{ 3, \begin{pmatrix} 1 \\ -1 \end{pmatrix} \right\}.$$

The continuous time model has thus a general solution of the form (3.152), namely

$$u(t) = A_1 \begin{pmatrix} 1 \\ 1 \end{pmatrix} \sin t + B_1 \begin{pmatrix} 1 \\ 1 \end{pmatrix} \cos t$$

$$+ A_2 \begin{pmatrix} 1 \\ -1 \end{pmatrix} \sin \sqrt{3} t + B_2 \begin{pmatrix} 1 \\ -1 \end{pmatrix} \cos \sqrt{3} t. \tag{4.82}$$

The system (4.80)-(4.81) is sampled with constant time intervals $h = 0.1$. The discrete time model is given by

$$100 \mathbf{M} \big(\mathbf{u}[k-1] - 2\mathbf{u}[\dot{k}] + \mathbf{u}[k+1] \big) + \mathbf{K} \mathbf{u}[k] = 0 \tag{4.83}$$

with the general solution

$$u(t) = \hat{A}_1 \begin{pmatrix} 1 \\ 1 \end{pmatrix} \sin \theta_1 k + \hat{B}_1 \begin{pmatrix} 1 \\ 1 \end{pmatrix} \cos \theta_1 k$$

$$+ \hat{A}_2 \begin{pmatrix} 1 \\ -1 \end{pmatrix} \sin \theta_2 k + \hat{B}_2 \begin{pmatrix} 1 \\ -1 \end{pmatrix} \cos \theta_2 k, \tag{4.84}$$

where

$$\theta_1 = \frac{\sqrt{0.04 - 0.0001}}{2 - 0.01} = 0.1004$$

and

$$\theta_2 = \frac{\sqrt{0.12 - 0.0009}}{2 - 0.03} = 0.1752.$$

Noting that $t = hk$ and that $\sqrt{3} = 1.732$ we find that the discrete time model approximates well the dynamics of the mass-spring system. ■

4.6. Finite Difference Model of an Eigenvalue Problem

Consider the axially vibrating rod of cross-sectional area $A(x)$, modulus of elasticity $E(x)$ and density $\rho(x)$, fixed at $x=0$ and free to vibrate at $x=L$, as shown in Figure 4.8.

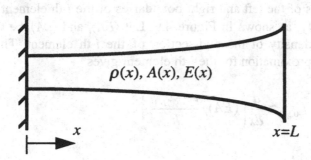

$$\rho(x),\ A(x),\ E(x)$$

$$x \qquad\qquad\qquad\qquad x=L$$

Figure 4.8. The non-uniform rod

The axial motion of the rod is governed by (3.199), namely,

$$\frac{\partial}{\partial x}\left(E(x)A(x)\frac{\partial u}{\partial x}\right) = \rho(x)A(x)\frac{\partial^2 u}{\partial t^2} \tag{4.85}$$

and the boundary conditions

$$\begin{cases} u(0,t) = 0 \\ \dfrac{\partial u(L,t)}{\partial x} = 0, \end{cases} \tag{4.86}$$

Harmonic motion assumption

$$u(x,t) = v(x)\sin\omega t , \tag{4.87}$$

leads to the associated eigenvalue problem

$$\begin{cases} \left(E(x)A(x)v'(x)\right)' + \lambda\rho(x)A(x)v(x) = 0, \quad \lambda = \omega^2, \\ v(0) = 0, \\ v'(L) = 0, \end{cases} \tag{4.88}$$

where primes denote derivatives with respect to x.

Depending on the functions $A(x)$, $E(x)$ and $\rho(x)$, the eigenvalue problem (4.88) may not be solvable in closed form. We may approximate some eigenvalues and eigenvectors of (4.88) by using finite differences as follows.

Partition the rod into n elements of equal length $h = L/n$ and denote the displacements of the left and right boundaries of the i-th element by v_{i-1} and v_i, respectively, as shown in Figure 4.9. Let $(EA)_i$ and $(\rho A)_i$ be the flexural rigidity and density of the mid-section of the i-th element. Then a finite difference approximation for the i-th element gives

$$\left(EAv'\right)'_{x=(i-0.5)h} \cong \frac{d}{dx}\left((EA)_i \frac{v_i - v_{i-1}}{h}\right), \tag{4.89}$$

and similarly

$$\begin{aligned}
\frac{d}{dx}\left((EA)_i \frac{v_i - v_{i-1}}{h}\right)_{x=ih} &\cong \frac{(EA)_{i+1}\dfrac{v_{i+1} - v_i}{h} - (EA)_i \dfrac{v_i - v_{i-1}}{h}}{h} \\
&= \frac{(EA)_i}{h^2} v_{i-1} - \frac{(EA)_i + (EA)_{i+1}}{h^2} v_i + \frac{(EA)_{i+1}}{h^2} v_{i+1}.
\end{aligned} \tag{4.90}$$

The finite difference scheme for the differential equation in (4.88) is therefore

$$\frac{(EA)_i}{h^2} v_{i-1} - \frac{(EA)_i + (EA)_{i+1}}{h^2} v_i + \frac{(EA)_{i+1}}{h^2} v_{i+1} + \lambda(\rho A)_i v_i = 0, \tag{4.91}$$

or

$$\begin{aligned}
-\frac{(EA)_i}{h} v_{i-1} &+ \frac{(EA)_i + (EA)_{i+1}}{h} v_i - \frac{(EA)_{i+1}}{h} v_{i+1} \\
&- \lambda h(\rho A)_i v_i = 0.
\end{aligned} \tag{4.92}$$

Define

$$k_i = \frac{(EA)_i}{h} \tag{4.93}$$

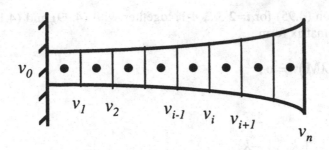

Figure 4.9. Finite difference mesh

and

$$m_i = h(\rho A)_i.$$ (4.94)

Then (4.92) can be written as

$$-k_i v_{i-1} + (k_i + k_{i+1})v_i - k_{i+1}v_{i+1} - \lambda m_i v_i = 0.$$ (4.95)

In terms of finite differences, the left boundary condition in (4.88) gives

$$v_0 = 0$$ (4.96)

and the right boundary implies that

$$\frac{v_{n+1} - v_n}{h} = 0,$$ (4.97)

or

$$v_{n+1} = v_n.$$ (4.98)

With (4.96), equation (4.95) gives for the first element, $i=1$

$$(k_1 + k_2)v_1 - k_2 v_2 - \lambda m_1 v_1 = 0.$$ (4.99)

With (4.98), equation (4.95) gives for the last element, $i=n$

$$-k_n v_{n-1} + (k_n + k_{n+1})v_n - k_{n+1}v_n - \lambda m_n v_n$$
$$= -k_n v_{n-1} + k_n v_n - \lambda m_n v_n = 0.$$ (4.100)

Equation (4.95) for $i=2,3,...,n-1$, together with (4.99) and (4.100) can be written in matrix form

$$(\mathbf{K} - \lambda\mathbf{M})\mathbf{v} = \mathbf{o}, \tag{4.101}$$

where

$$\mathbf{K} = \begin{bmatrix} k_1 + k_2 & -k_2 & & & \\ -k_2 & k_2 + k_3 & -k_3 & & \\ & -k_3 & k_3 + k_4 & -k_4 & \\ & & \ddots & \ddots & \ddots \\ & & & -k_n & k_n \end{bmatrix}, \tag{4.102}$$

$$\mathbf{M} = \begin{bmatrix} m_1 & & & & \\ & m_2 & & & \\ & & m_3 & & \\ & & & \ddots & \\ & & & & m_n \end{bmatrix}, \tag{4.103}$$

and

$$\mathbf{v} = (v_1, v_2, ..., v_n)^T. \tag{4.104}$$

Problem (4.101) is the well known generalised eigenvalue problem. It can be solved by standard routines (eg. *eig*(**K**,**M**) in Matlab). The eigenvalues of (4.101) approximate the square of the natural frequencies of the non-uniform rod. The eigenvectors of (4.101) are the approximations of the corresponding mode shapes, measured at $x=ih$, $i=1,2,...,n$.

This result has an interesting interpretation. We note that m_i in (4.94) is the mass of the i-th element. Also, k_i in (4.93) represents the stiffness of a rod of cross-sectional area A_i, length h and modulus of elasticity E_i. The model (4.101)-(4.104) thus describes an n degree-of-freedom mass-spring system of n masses, each equal to the mass of its associated rod element. They are connected by rods of length h, each with a stiffness corresponding to the stiffness of the appropriate rod element, as shown in Figure 4.10.

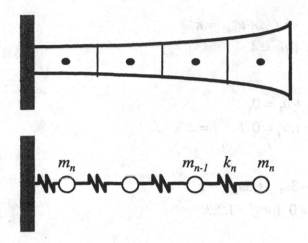

Figure 4.10. A lumped parameter model

The finite element model which has been developed mathematically in this section can thus be obtained simply by inspection. All we need to do is to concentrate the mass of each element in the element's center, and to connect the masses by springs with constants equal to the element stiffnesses, as shown above. Then the discrete model obtained can be analysed by the methods developed in Chapter 3. This model is called a *lumped parameter model*.

4.7 Exercises

1. In a manner similar to that developed in Section 4.2, determine a central finite difference scheme that approximates the first and second derivatives of $f(x)$ based on the slope of the parabola that collocates the three points $\{x - h, f(x - h)\}$, $\{x, f(x)\}$ and $\{x + h, f(x + h)\}$. Finf a bound the maximal error of the approximation and determine the order of the expected error for the two derivatives.

2. Determine the analytical solution of the following problems and compare the results with computer simulation for $k = 1, 2, ..., 10$.

   (a) $20u_k - 14u_{k+1} + 2u_{k+2} = 6$

   $u_1 = u_2 = 0.$

(b) $7u_k - 6u_{k+1} + u_{k+2} = k^2$

   $u_1 = 1, u_2 = 4.$

(c) $u_{k+5} + u_k = 0,$

   $u_1 = 1, u_j = 0$ for $j = 2,3,4,5.$

(d) $u_k - 3u_{k+1} + 3u_{k+2} - u_{k+3} = 1$

   $u_j = 0$ for $j = 1,2,3.$

3. Determine a first order realisation (ie. an equivalent system of difference equations of first order) for the system of difference equations

$$\begin{cases} x_k + 3x_{k+1} - 2y_{k-1} + 4z_{k+2} = 5 \\ x_{k+1} - 7y_{k+1} + 5z_{k+1} + 3z_{k+2} = 0 \\ 3x_k - 3y_{k-1} + 6z_{k+1} = 3. \end{cases}$$

4. Determine the analytical solution of

$$\begin{cases} x_k - 2y_k + 2x_{k+1} = 0 \\ y_k - 2x_k + 2y_{k+1} = 0 \end{cases}$$

with the initial conditions $x_1 = x_2 = 1$ by using the method developed in Section 3.2. Confirm the result by observation.

5. It was observed that for the mass shown in Figure 4.11, $x(0) = 0\,\text{m}$ and $x(0.1) = 0.2\,\text{m}$.

   (a) Determine a discrete time model with sampling rate $\Delta t = 0.1\,\text{s}$ for the system.

   (b) By using the discrete time model, approximate the position of the mass at the time $t = 1\,\text{s}$ and compare the approximation to the exact solution.

   (c) What is the exact and approximated final position of the mass?

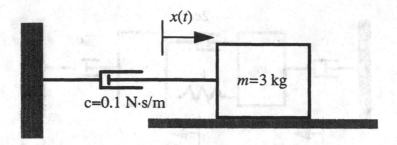

Figure 4.11. A mass-damper system

6. Consider the mass-spring-damper system shown in Figure 4.12.

   (*a*) Determine a discrete time model for the motion of the mass.

   (*b*) What is the maximal sampling rate Δt_{MAX} which ensures that the response obtained from the discrete time model is stable.

   (*c*) Using the sampling rate $\Delta t = 0.2 \Delta t_{MAX}$ simulate the response of the system in the time interval $0 \le t \le 5$ s.

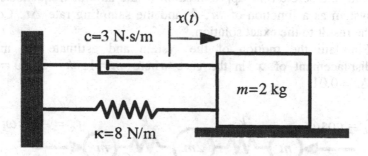

Figure 4.12. A mass-spring-damper system

7. For the two degree-of-freedom system shown in Figure 4.13 determine:

   (*a*) a discrete time model for $x(t)$ and $y(t)$,

   (*b*) the relation involving m, c, and k that guarantees unconditional stability of the discrete time model obtained in part (*a*), ie. the condition that ensures that the discrete model is stable for each sampling rate Δt, and

   (*c*) a discrete time approximation to the natural frequencies of the system, and compare the result to the exact solution.

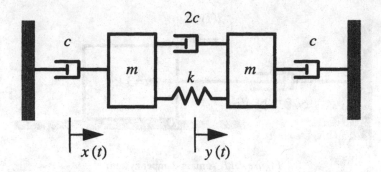

Figure 4.13. Two degree-of-freedom system

8. It was found that for the system shown in Figure 4.14

$$x(0) = y(0) = y(0.01) = z(0) = 0, \ x(0.01) = 0.1, \ z(0.01) = -0.1.$$

(*a*) Determine a discrete time model for $x(t)$, $y(t)$ and $z(t)$.

(*b*) Find a discrete time approximation to the natural frequencies of the system as a function of m, k and the sampling rate Δt. Compare the result to the exact solution.

(*c*) Simulate the motion of the system and estimate the maximal displacement of x in the case where $m = 1$, $k = 3$, $\omega = 1$ and $\Delta t = 0.01$.

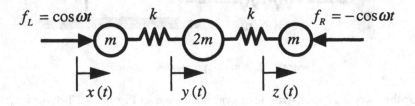

Figure 4.14. A three degree-of-freedom system

9. The particle of mass m shown in Figure 4.15 oscillates in a plane. Determine a non-linear discrete time model for the motion of the particle and simulate the motion for $0 \le t \le 5$ using sampling rate $\Delta t = 0.01\,\text{s}$, $k = 4\,\text{N/m}$, $m = 2\,\text{kg}$, and the conditions $x(0) = \theta(0) = 0$, $x(0.01) = 0.05\,\text{m}$, $\theta(0.01) = 0.02$ rad.

The system is a conservative system. Determine for each time step the total energy of the system and observe its variation, resulting from the discrete approximation of the motion.

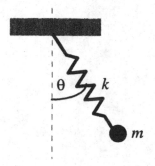

Figure 4.15. A flexible pendulum

10. Using 10 elements of equal length, determine a finite difference model for the axial vibrations of the uniform rod shown in Figure 4.16.

Approximate 10 of its damped natural frequencies for the case where $k = 4$ and $c = L = \rho = E = A = 1$.

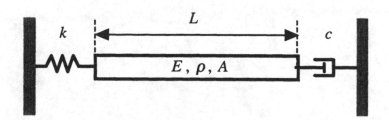

Figure 4.16. An axially vibrating rod

Bibliography

1. Batchelder, P.M., *An Introduction to Linear Difference Equations*, The Harvard University Press, Cambridge, Massachusetts, 1927.
2. Greenspan, D., *Discrete Models*, Addison-Wesley, London, 1973.
3. Lakshmikantham, V., and Trigiante, D., *Theory of Difference Equations: Numerical Methods and Applications*, Academic Press, Boston, 1988.

Bibliography

Chapter 5

NUMERICAL TECHNIQUES FOR MODEL NONLINEAR PDE's.

5.1 Introduction

The prototype for hyperbolic waves is often taken to be the wave equation

$$u_{tt} = c_0^2 \nabla^2 u, \tag{5.1}$$

although the equation

$$u_t + c_0 u_0 = 0 \tag{5.2}$$

is the simplest of all. Although classical problems lead to equation (5.1), many wave motions, e.g., flood waves, certain wave phenomena in chemical reactions, etc., have now been studied which do in fact lead to equation (5.2). The original formulations lead to nonlinear equations, just as in the classical problems, and the simplest nonlinear equation is

$$u_t + C(u)u_x = 0, \tag{5.3}$$

where the propagation speed $C(u)$ is a function of the local disturbance u.

The first order equation (5.3) is called quasi-linear as it is nonlinear in u but is linear in the derivatives u_t, u_x. The general nonlinear first order equation for $u(x,t)$ is any functional relationship between u, u_t and u_x.

Also corresponding to equation (5.3), shock waves appear as discontinuities in u.

The simplest equation to combine nonlinearity with diffusion is

$$u_t + C(u)u_x = vu_{xx}, \tag{5.4}$$

and great interest was shown in it when Hopf and Cole [1] discovered the transformation

$$u(x,t) = \frac{-2vu_x}{u}$$

which reduces it to the simple diffusion equation

$$u_t = vu_{xx}.$$

Hence the general solution can be obtained explicitly. Also equation (5.4) approximates the motion of a plane wave of small but finite amplitude. It takes into account both diffusion and convection, where

$v = $ diffusivity of sound,

$u = $ excess wavelet velocity,

$x = \bar{x} - c_0t = $ coordinate whose origin moves in the wave direction,

$c_0 = $ undisturbed speed of sound.

The special case of equation (5.4) with $C(u) = u$, namely,

$$u_t + uu_x = vu_{xx} \tag{5.5}$$

where $u = u(x,t)$ is some domain and v represents viscosity, is called Burgers' eqation. It is an important model equation and arises in the solution of many physical problems. Typical examples include such areas as gas dynamics, wave motions (e.g. flood waves, waves in glaciers, waves in traffic flow, wave phenomena in chemical reactions), shock waves, underwater acoustics, acoustic transmission in fog, and so on. Perhaps two of the key areas where it occurs most frequently are fluid dynamics (resulting from Navier-Stokes equation) and convection-diffusion.

In fluid dynamics the velocity field $\mathbf{v}$ and pressure field p for an incompressible Newtonian fluid with density ρ subjected to a body force $\mathbf{F}$ satisfy the equation

$$\frac{\partial \mathbf{v}}{\partial t} + (\mathbf{v} \cdot \nabla)\mathbf{v} = -\frac{1}{\rho}\nabla p + \mathbf{F} + \nu\nabla^2\mathbf{v} \tag{5.6}$$

and the continuity equation

$$\nabla \cdot \mathbf{v} = 0. \tag{5.7}$$

Clearly problems where $p = 0$ and $\mathbf{F} = 0$ can be reduced to the model equation

$$\frac{\partial \mathbf{v}}{\partial t} + (\mathbf{v} \cdot \nabla)\mathbf{v} = \nu\nabla^2\mathbf{v} \tag{5.8}$$

In problems involving convection and diffusion the governing equation is

$$D\nabla^2 C - \mathbf{v} \cdot \nabla C = \frac{\partial C}{\partial t} \tag{5.9}$$

where C (mass/volume) is the concentration of the solute, $\mathbf{v}$ is the liquid velocity and D (length^2/time) is the diffusion coefficient. The first term is referred to as the diffusion term and the second term the convection term.

In a remarkable series of papers from 1939 to 1965 Burgers investigated various aspects of turbulence and developed a mathematical model illustrating the theory and also studied statistical and spectral aspects of the equation and related systems of equations. Cole [1] studied the general properties of the equation and outlined some of its various applications. He pointed out that it shows the typical features of shock wave theory, a nonlinear term tending to steepen the wave fronts and produce complete dissipation and a viscous term of higher order which prevents formation of actual discontinuities which tends to diffuse any differences in velocity. Burgers' equation is related to turbulence theory as a mathematical model and this is largely due to its similarity to the Navier-Stokes equation. Both contain nonlinear terms of the types: unknown function multiplied by a first derivative and both contain higher order terms multiplied by a small parameter.

More recently, many author have used a variety of numerical techniques in attempting to solve the eqation particularly for small values of viscosity v which correspond to steep fronts in the propagation of dynamic waveforms.

At this stage it is important to explain why this Chapter has been devoted to Burgers' equation. Firstly, as already explained Burgers' equation is a useful model for many physically interesting problems, particularly those of a fluid-flow nature, in which either shocks or viscous dissipation is significant in part of the region. For many combinations of boundary and initial conditions, an exact solution of Burgers' equation is readily available. Hence approximations to physically more complex problems are often sought so that the exact solution of Burgers' equation can be exploited.

Burgers' equation is probably one of the simplest nonlinear partial differential equations for which it is possible to obtain an exact solution. Also, depending on the magnitude of the various terms in the equation, it behaves as an elliptic, parabolic or hyperbolic partial differential equation. Hence it has been widely used as a model equation for testing and comparing computational techniques.

This Chapter goes on to summarise a number of different numerical approaches to Burgers' equation which have been used by Caldwell et al. with a view to drawing important conclusions on the methods used and the accuracy of results obtained. It is important to realise that such numerical and computational techniques could equally well be applied more generally to other similar nonlinear partial differential equations arising from the mathematical modelling of physical systems. In order to avoid unnecessary detail only the key points of the different numerical approaches will be presented as full details can be found in the individual references.

In the first case, finite-difference methods can be shown to be successful and accurate for large viscosity v which corresponds to small Reynolds number Re. The method of lines involving Finite Fourier series adopted by Caldwell and Wanless [2] produces a set of ODE's which can be solved for the amplitudes of the sine terms. In this way we have a neat alternative to the approach used by Sincovec and Madsen [3] who use a Runge-Kutta technique to solve the system of ODE's directly from the discretised equations.

Caldwell [4] demonstrates the importance of Fourier transform methods for certain types of problem, e.g. problems where the initial condition is in the form of a step function. Errors using finite difference methods can be reduced by developing a numerical procedure where the space derivatives are computed with very high accuracy using Fourier transform methods.

Finite difference methods prove to be inaccurate for large Reynolds number. For this case a piecewise polynomial approximation (i.e. finite element) was adopted by Caldwell, Wanless and Cook [5] and Caldwell and Smith [6]. Such a method is demonstrated for the case of fixed nodes using cubic polynomials for the shape functions. Improvements can be obtained by choosing the size of the elements to take into account the nature of the solution. The aim is to 'chase the peak' by altering the size of the elements at each stage using information from the previous step. – refer to Caldwell, Wanless and Cook [7]. Possible advantages over Finite Element techniques in using cubic splines are discussed by Caldwell [8] and certain conclusions are drawn.

Caldwell, Wanless and Burrows [9] give an indication of how complementary variational principles can be applied to non-linear equations and, in particular, to Burgers' equation. More detailed discussions of this method as applied to the steady-state form of Burgers' equation are given by Saunders, Caldwell and Wanless [10].

5.2 Finite Difference and Fourier Methods

The exact solution of Burgers' equation (5.5) under the boundary conditions

$$u(0,t) = u(1,t) = 0, \quad t > 0 \tag{5.10}$$

and the initial condition

$$u(x,0) = f(x), \quad 0 \leq x \leq 1 \tag{5.11}$$

is given by (see Cole [1]), namely

$$u(x,t) = \frac{2\pi v \sum_{n=1}^{\infty} nA_n \sin n\pi x \exp(-n^2\pi^2 vt)}{A_0 + \sum_{n=1}^{\infty} A_n \cos n\pi x \exp(-n^2\pi^2 vt)}, \tag{5.12}$$

where

$$A_n = 2\int_0^1 \cos n\pi x \exp\left\{-\frac{1}{2v}\int_0^x f(x')dx'\right\}dx, \quad (n=1,2,3...) \tag{5.13}$$

$$A_0 = \int_0^1 \exp\left\{-\frac{1}{2v}\int_0^x f(x')dx'\right\}dx. \tag{5.14}$$

For the special case $f(x) = \sin\pi x$ considered later, equation (5.12) becomes

$$u(x,t) = \frac{4\pi v \sum_{n=1}^{\infty} nI_n \sin n\pi x \exp(-n^2\pi^2 vt)}{I_0 + 2\sum_{n=1}^{\infty} I_n \cos n\pi x \exp(-n^2\pi^2 vt)} \tag{5.15}$$

where the modified Bessel functions are

$$I_n = \int_0^1 \cos n\pi x \exp\left(\frac{\cos n\pi x}{2\pi v}\right)dx, \quad (n=1,2,3,...) \tag{5.16}$$

$$I_0 = \int_0^1 \exp\left(\frac{\cos\pi x}{2\pi v}\right)dx. \tag{5.17}$$

For relatively large values of viscosity v (e.g. $v=1$) which correspond to small Reynolds number Re, the coefficients I_n ($n=1,2,3,...$) drop off very rapidly and hence high accuracy can be achieved by taking only a few terms in the series in (5.15). However, for large Reynolds number Re > 10 the convergence of the series is slow in most cases and so we must consider numerical approaches such as finite difference methods.

Typical finite difference methods which can easily be applied include:

(a) <u>Explicit scheme</u>

$$u_{i,j+1} = vr(u_{i+1,j} - 2u_{i,j} + u_{i-1,j}) + \left\{1 - \frac{rh}{2}(u_{i+1,j} - u_{i-1,j})\right\}u_{i,j} \tag{5.18}$$

where $r = k/h^2 = \delta t/(\delta x)^2$. In this case the nonlinearity is easily handled but the major drawback is that $0 < r \le 1/2$.

(b) Implicit scheme

$$u_{i,j+1} = u_{i,j} + vr(u_{i+1,j+1} - 2u_{i,j+1} + u_{i-1,j+1}) - \frac{rh}{2}(u_{i+1,j+1} - u_{i-1,j+1})u_{i,j+1}. \quad (5.19)$$

The unknown $u_{i,j+1}$ is now involved in a nonlinear expression and therefore must be solved iteratively e.g. Gauss-Seidel method or S.O.R.

(c) Crank-Nicolson method

$$u_{i,j+1} = u_{i,j} + \frac{vr}{2}(u_{i+1,j+1} - 2u_{i,j+1} + u_{i-1,j+1} + u_{i+1,j} - 2u_{i,j} + u_{i-1,j})$$

$$- \frac{rh}{4}\{(u_{i+1,j+1} - u_{i-1,j+1})u_{i,j} + (u_{i+1,j} - u_{i-1,j})u_{i,j+1}\}. \quad (5.20)$$

This is nonlinear in $u_{i,j+1}$. However, we can use the approximation

$$uu_x \cong \frac{1}{2}\{u(t + \delta t)u_x(t) + u(t)u_x(t + \delta)\},$$

where $u(t)$ and $u_x(t)$ are known. This results in a linear system of the form

$$\mathbf{A}(t)\mathbf{U}(t + \delta t) = \mathbf{B}(t)\mathbf{U}(t),$$

where $\mathbf{A}(t)$ and $\mathbf{B}(t)$ are tridiagonal matrices. Hence

$$\mathbf{U}(t + \delta t) = \mathbf{A}^{-1}(t)\mathbf{B}(t)\mathbf{U}(t)$$

The above standard schemes have been used with success on Burgers' equation for a typical test case $f(x) = \sin \pi x$ and prove accurate for large v.

In an attempt to reduce errors in the above methods, a numerical procedure can be developed where the space derivatives are computed with very high accuracy by means of Fourier transform methods. Using this approach a forward marching problem involves discrete time steps but space derivatives are accurate within the limit to which a distribution can be defined on a finite set of meshpoints. This is particularly useful for problems where the initial condition is in the form of a step function. For Burgers' equation we advance the solution by accounting for both the convective and diffusive terms simultaneously and this leads to considerable improvement in accuracy.

In this method $u(t+\Delta t)$ is computed from $u(t)$ using

$$u(t+\Delta t) = u(t) + \Delta t u_t + \frac{\Delta t^2}{2!} u_{tt} + \frac{\Delta t^3}{3!} u_{ttt} \tag{5.21}$$

and the time derivatives are computed using

$$u_t = -u u_x + v u_{xx} \tag{5.22}$$

$$u_{tt} = -u_t u_x - u u_{xt} + v u_{xxt} \tag{5.23}$$

$$u_{ttt} = -u_{tt} u_x - 2 u_t u_{xt} - u u_{xtt} + v u_{xxtt} \tag{5.24}$$

The x-derivative terms are computed by Fourier methods over the chosen solution domain D and the solution is advanced by means of the Fast Fourier Transform (FFT) method. The principal domain

$$D = \{x : 0 \le x \le L\} \tag{5.25}$$

is partitioned into two sub-domains D_1 and D_2 such that

$$D = D_1 + D_2.$$

Values of u over D_1 are fixed and kept constant throughout the entire computation and new u values are computed over the domain D_2. The purpose of D_1 is to provide a smooth transition between the two end-points of D_2 and to ensure periodicity over D.

As a test example $f(x)$, D_1 and D are taken to be

$$f(x) = \begin{cases} 0 & 0 \le x < 0.1 \\ [1 - \cos\{(x - 0.1)\pi / 0.3\}] / 2 & 0.1 \le x < 0.4 \\ 1 & 0.4 \le x < 0.6 \end{cases} \tag{5.26}$$

$$D_1 = \{x : 0 \le x \le 0.6\} \tag{5.27}$$

$$D = \{x : 0 \le x \le 2.4\} \tag{5.28}$$

and solutions are obtained using intervals $\Delta t = 0.001$, $\Delta x = 0.01$ for the case $v = 0.005$ (i.e. Re $= 200$). By considering the evolution of the numerical solution, the results show that the maximum error is less than 0.01% of the maximum value u_1.

5.3 Method of Lines and Fourier Approach

First of all, consider the solution of Burgers' equation analytically using the method of lines. The notation used is $u(x, t_r) = u(x, rk) = u_r(x)$ and u_t is approximated by $(u_{r+1}(x) - u_r(x)) / k$. This means that equation (5.5) can be approximated by

$$\frac{u_{r+1} - u_r}{k} = v \frac{d^2 u_{r+1}}{dx^2} - u_{r+1} \frac{du_{r+1}}{dx}. \tag{5.29}$$

Hence

$$\frac{d^2 u_{r+1}}{dx^2} - \frac{1}{vk} u_{r+1} = -\frac{1}{vk} u_r + \frac{1}{v} u_{r+1} \frac{du_{r+1}}{dx}. \quad (r = 0, 1, 2, \ldots) \tag{5.30}$$

Taking $r = 0$ and letting $B = 1 / vk$ gives

$$\frac{d^2 u_1}{dx^2} - B u_1 = -B u_0 + \frac{1}{v} u_1 \frac{du_1}{dx} \tag{5.31}$$

A first order solution is obtained by neglecting the nonlinear term on the right-hand side and using the initial condition $u_0 = \sin \pi x$. The general solution is

$$u_1 = ae^{\sqrt{B}x} + be^{-\sqrt{B}x} + \left\{B/(B+\pi^2)\right\}\sin \pi x \qquad (5.32)$$

where a and b are constants to be determined.

The boundary conditions (5.10) lead to the first order approximation

$$u_1 = \left\{B/(B+\pi^2)\right\}\sin \pi x \qquad (5.33)$$

Now iterate by explicitly working out the correction term $(u_1/v)du_1/dx$ using the linear u_1 and the equation resolved. In this way the correction term can be approximated by $(\pi/2v)\sin 2\pi x$. This introduces another term into the particular integral, namely $-\{\pi/2(4\pi^2+B)\}\sin 2\pi x$. Therefore, to second order the solution is

$$u_1 = \frac{B}{B+\pi^2}\sin \pi x - \frac{\pi}{2(4\pi^2+B)}\sin 2\pi x . \qquad (5.34)$$

This process can be repeated to give the approximate solution at higher time steps.

A more systematic approach is to use the method of lines involving finite Fourier series. The nonlinear terms will generate a solution different from the initial condition input.

Setting $v = 1/2A$ in equation (5.30) gives

$$\frac{d^2 u_{r+1}}{dx^2} - 2Au_{r+1}\frac{du_{r+1}}{dx} - Bu_{r+1} = -Bu_r . \quad (r = 0, 1, 2, \ldots) \qquad (5.35)$$

Imposing the boundary conditions $u_r(0) = u_r(1) = 0$ and the initial condition $u_0(x) = f(x)$ or some approximation to $f(x)$, leads to a set of nonlinear equations because of the nonlinear term. To the same order of accuracy, equation (5.35) may be replaced by the linear system

$$\frac{d^2 u_{r+1}}{dx^2} - Au_{r+1}\frac{du_r}{dx} - Au_r\frac{du_{r+1}}{dx} - Bu_{r+1} = -Bu_r \quad (r = 0, 1, 2, \ldots) \quad (5.36)$$

As in the case of equation (5.31), the solution of equation (5.36) will require only the determination of the particular integral, because of the boundary conditions.

Assuming the particular integral can be expressed in the form

$$u_r(x) = \sum_{j=1}^{N} \alpha_j \sin j\pi x \qquad (\alpha \text{ known})$$

$$u_{r+1}(x) = \sum_{k=1}^{N} \beta_k \sin k\pi x \qquad (\beta \text{ unknown})$$

(5.37)

then substitution in equation (5.36) gives

$$\sum_{k=1}^{N}(B+k^2\pi^2)\beta_k \sin k\pi x + \frac{1}{2}A\pi \sum_{j=1}^{N}\sum_{k=1}^{N} j\alpha_j \beta_k \{\sin(k+j)\pi x +$$

$$\sin(k-j)\pi x\} + \frac{1}{2}A\pi \sum_{j=1}^{N}\sum_{k=1}^{N} k\alpha_j \beta_k \{\sin(k+j)\pi x + \sin(j-k)\pi x\} \quad (5.38)$$

$$= B\sum_{j=1}^{N}\alpha_j \sin j\pi x.$$

Equating coefficients of $\sin i\pi x$ in equation (5.38) gives

$$(B+i^2\pi^2)\beta_i + \frac{1}{2}A\pi i\left\{\sum_{j=1}^{i-1}\alpha_{i-j}\beta_j - \sum_{j=i+1}^{N}\alpha_{j-i}\beta_j - \sum_{j=1}^{N}\alpha_{i+j}\beta_j\right\} = B\alpha_i .$$

(5.39)

$$(i = 1,2,3,\ldots,N)$$

This results in a system of N simultaneous linear equations of the form

$$\mathbf{C}\beta = \alpha$$

(5.40)

where the (i, j) coefficient of the matrix $\mathbf{C}$ is given by

$$\mathbf{C}_{ij} = \frac{1}{B}\left\{(B+i^2\pi^2)\delta_{ij} + \frac{1}{2}A\pi i(\alpha_{i-j} - \alpha_{j-i} - \alpha_{i+j})\right\}$$

and

$$\alpha_k = 0 \qquad \text{if} \quad k \leq 0$$

$$\delta_{ij} = \begin{cases} 1 & i = j \\ 0 & i \neq j \end{cases}$$

Hence the coefficient β_i, $(i = 1,2,3,\ldots,N)$ can be computed for any order N.

5.4 Finite Element Methods

The methods described in Sections 5.2 and 5.3 produce accurate results for large v. However, problems arise for small values of v which correspond to steep fronts in the propagation of dynamic waveforms. This indicates that for small v a piecewise polynomial approximation (i.e. Finite Element) should be considered where the size of the elements is chosen to take into account the nature of the solution.

Fixed nodes. This section describes the extension of the finite element approach proposed by Caldwell et al [5] to the general case of n elements (fixed).

On setting $Y = u_r$, and $Z = u_{r+1}$ in equation (5.36), this leads to the equation

$$Z'' - AYZ' - (AY' + B)Z + BY = 0 \tag{5.41}$$

subject to the specified initial and boundary conditions.

The residual R may be defined as

$$R = \sum_{i=1}^{n-1} \int_{a_i}^{a_{i+1}} \left\{ Z'' - AYZ' - (AY' + B)Z + BY \right\}^2 dx , \tag{5.42}$$

where

$$R = \sum_{i=1}^{n} R_i$$

and R_i, $i = 1,2,\ldots,n$, refer to the residuals in each of the n elements.

Denoting the solution at time t by Y and at time $(t+\delta t)$ by Z, this gives

$$Y = \sum_{i=1}^{n} Y^{(i)}, \qquad Z = \sum_{i=1}^{n} Z^{(i)} \qquad\qquad (5.43)$$

where $Y^{(i)}$, $Z^{(i)}$ are defined on the ith element. Both Y and Z can be expressed in terms of shape functions as follows

$$Y^{(i)} = N_{4i-3} y_i + N_{4i-2} y_i' + N_{4i-1} y_{i+1} + N_{4i} y_{i+1}' \qquad a_i < x < a_{i+1} \qquad (5.44)$$

$$Z^{(i)} = N_{4i-3} z_i + N_{4i-2} z_i' + N_{4i-1} z_{i+1} + N_{4i} z_{i+1}' \qquad b_i < x < b_{i+1} \qquad (5.45)$$

where y_i, y_i' are the values of u and the gradient at the two 'end points' of the ith element.

The shape functions for the ith element can be determined by assuming each is equal to a general cubic in x with unknown coefficients and then evaluating each shape function and its derivative at both ends of the ith element. This results in four equations which can be evaluated in terms of b_i and b_{i+1}.

The residuals R_i all satisfy the equation

$$R_i = \int_{b_i}^{b_{i+1}} \left\{ Z^{(i)''} - AYZ^{(i)'} - (AY'+B)Z^{(i)} + BY \right\}^2 dx \qquad (5.46)$$

and the following stationary conditions are sought

$$\frac{\partial R}{\partial z_j} = \frac{\partial R}{\partial z_j'} = 0 \qquad j = 1, 2, 3, \ldots, n+1 \qquad (5.47)$$

leading to a system of $2n+2$ equations in z_j and z_i'.

Now for the ith element the conditions

$$\frac{\partial R}{\partial z_i} = \frac{\partial R}{\partial z_i'} = \frac{\partial R}{\partial z_{i+1}} = \frac{\partial R}{\partial z_{i+1}'} = 0 \qquad (5.48)$$

must be satisfied. In matrix form, this results in the elemental system of equations

$$K_i z_i = -B d_i .$$ (5.49)

For the case of n elements this results in the global system of equations

$$\mathbf{Kz} = -\mathbf{Bd}$$ (5.50)

of order $2n+2$. This system can then be solved to give $\mathbf{z}$ and hence values of z and z' at each nodal point. It should be noted that z_1 and z_{n+1} will be specified at the end points.

Moving nodes. Caldwell et al [5] have demonstrated how the finite element method can be used to solve Burgers' equation by using moving nodes. The aim is to the 'chase the peak' by altering the size of the element at each stage using information from the previous step. This involves moving the nodes from the initial mesh configuration to regions where the residuals are relatively high. This has been described for the case of two elements. However, this work is easily extended to the general case of n elements. The simple bisection technique could be used to search for the turning point in the solution. The remaining internal nodes could then be located around it and the new node set $\mathbf{x}$ could be used as a starting point for the next time step. This procedure could then be repeated a number of times.

The following algorithm describes this process in more detail and applies if n is even and $n \geq 6$, i.e. at least five interior nodes:

(a) Start with evenly spaced nodes. Define initial node set $\mathbf{x}^0$. Set $j = 0$.
(b) Solve the system $\mathbf{Kz}^j = -\mathbf{Bd}$ for $\mathbf{z}^j$.
(c) If $|z_i'| < \varepsilon$ then go to (h). This applies for any i, $i = 2,3,\ldots,n$.
(d) Test to find an interval which contains the turning point, i.e. if $z_i^j z_{i+1}^j < 0$ then the turning point lies within the interval (x_i^j, x_{i+1}^j).
(e) Move the central node to the centre of this interval, i.e.

$$x_{n/2+1}^{j+1} = x_i^j + \frac{1}{2}\left(x_{i+1}^j - x_i^j\right)$$

(f) Equally space the remaining interior nodes, i.e.

$$x_k^{j+1} = \frac{2(k-1)}{n-2} x_i^j; \quad k = 2,3,\ldots,\frac{n}{2}-1$$

and

$$x_i^{j+1} = \frac{(2l-n-4)}{n-2}\left(1-x_{i+1}^j\right)+ x_i^j ; \quad l = \frac{n}{2}+3, \frac{n}{2}+4,..., n .$$

Thus the new node set $\mathbf{x}^{j+1}$ is defined.
(g) Set $j = j+1$. Go to (b).
(h) The turning point has now been determined. Place the central node at this point if it is not already positioned there. Distribute the other nodes around this central node (at $x_{n/2+1}^*$), i.e.

$$x_m = \frac{1}{2}\left(x_{n/2+1} - x_{m-1}\right), \qquad m = 2,3,...,\frac{n}{2}$$

$$x_p = \frac{1}{2}\left(x_{p+1} - x_{n/2+1}\right), \qquad p = n, n-1,...,\frac{n}{2}.$$

(i) Set $t = t+\delta t$. Use this new node set as $\mathbf{x}^0$ for the next time step. Stop if t equals the final time. Otherwise go to (b).

Cubic splines. In the numerical solution of PDE's cubic splines are superior in some cases to Finite Element techniques in that the normal disadvantages of high computer cost and complex problem formulation are removed. In general, there are two main advantages:
(1) the governing matrix system obtained is always tridiagonal which permits the use of an efficient matrix inversion procedure.
(2) the matrix system can be reduced to a scalar system of equations containing values of the function, first and second derivatives at the node points while maintaining a tridiagonal formulation.

However, in applying this method of splines to Burgers' equation, we find that it is only satisfactory for $v = 1$ or close to 1. For $v < 0.1$ the results are inaccurate. Values of Reynolds number Re from 10 up to 10^{25} (i.e. corresponding to values of v from 10^{-1} down to 10^{-25}) have been tested to check the claim of Wang and Kahawita [12] that when Re 'becomes large' splines again become computationally viable. However, this view is not supported by the results obtained.

5.5 Variational-Iterative Scheme

Complementary variational principles have been introduced by Kato [12] and others up to Walpole [13] for the equation

$$T\phi = f \tag{5.51}$$

in a suitable chosen and real Hilbert space H where T is a completely continuous self-adjoint operator and f is some function belonging to H. These principles provide upper and lower bounds for $\langle \phi / f \rangle$ where $\langle\ \rangle$ denotes the inner product of H. Arthurs [14] has also introduced complementary variational principles for the solution of equations of the form

$$T\phi = f(\phi) \tag{5.52}$$

where f may be a nonlinear function of ϕ.

In later work Burrows and Perks [15] have extended their linear theory to deal with the nonlinear equation (5.52). A variational-iterative scheme is used to deal with the problems provided by equation (5.52) so that the simple linear theory is applied to a sequence of equations, the solutions of which converge to the solutions of equation (5.52).

By considering the functional

$$J(\Delta, \phi) = \langle \phi / T\phi \rangle - 2\langle \phi / f \rangle + \Delta \langle T\phi - f / T\phi - f \rangle \tag{5.53}$$

where Δ is a real constant, Burrows and Perks demonstrate that the real quantity

$$S = \min_{\psi \in H} \left\{ J(\Delta_1, \omega_p) - J(\Delta_2, \psi) \right\} \tag{5.54}$$

provides a measure of the convergence criteria. Here ψ denotes the exact solution; Δ_1, Δ_2 refer to the minimum and maximum principles respectively and ω_p denotes the limit of the sequence of trial functions $\{\omega_{n,p}\}$ containing p variational parameters for each iterate.

In applying this work to the solution of nonlinear equations, the basic idea is to attempt to rearrange the nonlinear equation

$$A\phi = f(\phi) \tag{5.55}$$

into the form (5.52) where T is self-adjoint on the space considered and such that T has a discrete spectrum. Then iteration with the sequence of equations

$$T\psi_{n+1} = f(\Phi_{n+1}) \tag{5.56}$$

gives Φ_{n+1} as a variational approximation to ψ_{n+1}. Under certain conditions the sequence $\{\Phi_{n+1}\}$ will converge to ϕ.

In the first instance these principles have been tested out on the steady state version of Burgers' equation under the boundary conditions

$$u(0) = 0, \qquad u(\infty) = -2v \tag{5.57}$$

which has the exact solution

$$u = -2v \tanh x.$$

As a result of the success obtained this variational-iterative scheme can be applied to the full Burgers' equation (5.5) under the boundary conditions (5.10) and initial condition

$$u(x,0) = \sin \pi x, \qquad 0 \le x \le 1. \tag{5.58}$$

Since $u(x,t) \to 0$ as $t \to \infty$ it is appropriate to assume

$$u(x,t) \underset{t \to \infty}{\cong} \exp(-\alpha t). \qquad (\alpha > 0)$$

With this assumption it is best to substitute $T = 1 - \exp(-t)$. This leads to

$$(1-T)u_T + uu_x = vv_{xx} \tag{5.59}$$

inside the rectangle $0 \le x \le 1$, $0 \le T \le 1$.

Equation (5.59) can be rewritten as

$$u_{xx} + u_{TT} = (1+v)u_{xx} + u_{TT} - uu_x - (1-T)u_T$$

which has the form

$$\nabla^2 u = F(u). \tag{5.60}$$

Denoting the interior of the rectangle by V with boundary δV, Green's formula gives

$$\int_V (\phi, \nabla^2 u_1 - u_1 \nabla^2 \phi_1) dV = \int_{\delta V} (\phi_1, \nabla u_1 - u_1 \nabla \phi_1) \cdot \hat{n}\, dS$$

which means that ∇^2 is self-adjoint if the space is such that $\phi_1 = u_1 = 0$ on δV. This is obtained by writing

$$u = f(x,T) + \phi \tag{5.61}$$

where $f(x,T)$ obeys the boundary conditions of the problem and hence $\phi = 0$ on δV. A suitable choice for $f(x,T)$ is

$$f(x,T) = (1-T)\sin \pi x.$$

Hence equation (5.60) becomes

$$\nabla^2 \phi = F(u) - \nabla^2 f = G(\phi)$$

which means that $T = \nabla^2$. The diserete spectrum of ∇^2 on a rectangle is well known with zero boundary conditions.

In this way Burgers' equation has been reduced to a suitable form which enables the variational-iterative scheme to be applied.

5.6 Discussion of Numerical Results and Conclusions

Results have been obtained by applying these numerical techniques to the important model nonlinear PDE, namely, Burgers' equation (5.5), under the boundary conditions (5.10) and initial condition (5.58). Where possible these results have been compared with the analytical results. Because of convergence problems in the analytical solution this has not always been possible for small viscosity v.

It has been found that for large v (say $v = 1$) the finite difference (FD) results (particularly the explicit method described in Section 5.2) are more accurate than the finite element (FE) results. Although there is good agreement between the FE results with fixed nodes with $n = 4$ and 20, these results do not even agree with the exact results to 2 decimal places. Little improvement is obtained by using the FE method with moving nodes (as described in Section 5.4). Accuracy in excess of that obtained by the FD methods is achieved by using the Fourier series approach (as described in Section 5.2). A comparison of results is presented in Tables 5.1 and 5.2 for the cases $t = 0.01$ and $t = 0.05$ respectively for $v = 1$ using $h = \delta x = 0.25$ and $k = \delta t = 0.01$.

Method	$u(x, 0.01)$		
	$x = 0.25$	$x = 0.50$	$x = 0.75$
FD (explicit)	0.6267	0.9063	0.6550
FD (implict)	0.6377	0.9141	0.6556
Fourier ($N = 4$)	0.6302	0.9068	0.6528
FE ($n = 4$)	0.6333	0.9100	0.6539
FE ($n = 20$)	0.6335	0.9101	0.6539
Exact	0.6290	0.9057	0.6524

Table 5.1 Comparison of solutions $u(x, t)$ of Burgers' equation for the case $t = 0.01$, $v = 1$.

Method	$u(x, 0.05)$		
	$x = 0.25$	$x = 0.50$	$x = 0.75$
FD (explicit)	0.4099	0.6100	0.4556
FD (implict)	0.4339	0.6380	0.4702
Fourier ($N = 4$)	0.4157	0.6128	0.4528
FE ($n = 4$)	0.4231	0.6233	0.4601
FE ($n = 20$)	0.4237	0.6239	0.4604
Exact	0.4131	0.6091	0.4502

Table 5.2 Comparison of solutions $u(x, t)$ of Burgers' equation for the case $t = 0.05$, $v = 1$.

Fourier transform methods have proved successful and accurate for certain types of problem e.g., where the initial condition is in the form of a step function. In the test case mentioned in Section 5.2 (see equations (5.26–5.28)) accuracy of the order of 0.01% is obtained for $v = 10^{-2}$ and the accuracy improves with increasing v. The high accuracy of this method is attributed to its small truncation errors which are $O(\Delta t^4)$. More accurate representations of such steep profile requires higher wave numbers than those that can be supported by the computational mesh used.

For small v the FE results are much superior to the FD and Fourier series results. For example, as discussed by Caldwell and Smith [6], for $v = 0.01$ the fixed node $n = 20$ results are correct to 4 decimal places and the $n = 4$ results are correct to 3 decimal places in most cases over the range $0 \le t \le 0.25$ whereas the FD and Fourier results are completely inaccurate. Further improvement is obtained by using moving nodes. For very small v

(say $v = 10^{-4}$) the agreement is even better, namely 7 decimal places for fixed node $n = 20$ and 4 decimal places for $n = 4$. Higher accuracy is achieved for smaller values of v.

The inaccurate FD results are explained by the fact that as v decreases from 10^{-2} to 10^{-4} a disturbance appears at $x = 0.5$ for small t and moves towards $x = 1$ with steepened front as t increases. After the disturbance reaches a maximum near $x = 1$ for some time t, all values of u tend to decrease in a uniform manner. As v decreases more disturbances appear giving a ripple effect which becomes more exaggerated and decays more slowly as v tends to 10^{-4}.

Only very limited improvement in the explicit and implicit FD schemes (5.18) and (5.19) is achieved by using the Crank-Nicolson method given in (5.20). Attempting to overcome the nonlinearity problem by using a predictor-corrector modification of the Crank-Nicolson method described by Ames [16] still gives great difficulty in obtaining computed solutions which tend to the smooth asymptotic profile of Cole [1] particularly for small v. The Fourier series approach works well for values of v down to 0.1 where the analytical formula is not particularly helpful because of slow convergence. This method also avoids the stability problems associated with the traditional FD techniques. Many of these difficulties are overcome by using the fixed node FE approach (as described in Section 5.4) for the case of small viscosity v. Further improvement is obtained by 'chasing the peak' of the developing wave through adjustment of the element sizes.

To date the variational-iterative scheme has been applied to the steady state version of Burgers' equation. Variational approximations of increasing accuracy have been generated and the method has been shown to converge by examining variations of measures of convergence. These encouraging results give confidence in the applications of complementary variational principles to Burgers' equation itself and this work is in progress.

It is important to point out that the above methods are not restricted to Burgers' equation but can be applied to other similar nonlinear PDE's arising in the modelling process. As noted earlier, Burgers' equation has been taken as a test example because it is one of the simplest nonlinear PDE's for which an exact solution exists. Two other obvious important nonlinear PDE's which should be mentioned are :

(1) Korteweg-de Vries equation, namely

$$u_t + uu_x + \mu u_{xxx} = 0 . \tag{5.62}$$

This equation models many physical systems ranging from water waves and lattice waves to plasma waves.

(2) a combination of Korteweg-de Vries equation and Burgers' equation, namely

$$u_t + uu_x + \mu u_{xxx} = vu_{xx} . \tag{5.63}$$

This equation models some classes of nonlinear dispersive systems with dissipation. Physical considerations require that the dissipative parameter v must always be positive, while the dispersive parameter μ may be either positive or negative.

5.7 Exercises

1. Show that the exact solution of Burgers' equation

$$u_t + uu_x = vu_{xx}$$

under the boundary conditions

$$u(0,t) = u(1,t) = 0, \quad t > 0$$

and initial condition

$$u(x,0) = f(x), \qquad 0 \le x \le 1$$

is given by

$$u(x,t) = \frac{2\pi v \sum_{m=1}^{\infty} mA_m \sin m\pi x \exp(-m^2 v \pi^2 t)}{A_0 + \sum_{m=1}^{\infty} A_m \cos m\pi x \exp(-m^2 v \pi^2 t)},$$

where

$$A_m = 2\int_0^1 \cos m\pi x \exp\left(-\frac{1}{2v}\int_0^x f(x')dx'\right)dx \qquad (m = 1,2,3,...)$$

$$A_0 = \int_0^1 \exp\left(-\frac{1}{2v}\int_0^x f(x')dx'\right)dx \,.$$

2. Find the solution $u(x,t)$ in Exercise 1 for the case $f(x) = \sin \pi x$. Show that, provided v is sufficiently large, a reasonable approximation for $u(x,t)$ is given by

$$u(x,t) = \frac{\sin \pi x \exp(-v\pi^2 t)}{1+(1/2\pi v)\cos \pi x \exp(-v\pi^2 t)} \qquad (t > 0)$$

3. Find the solution $u(x,t)$ in Exercise 1 for the special case $f(x) = \sin \pi x$ and $v = 1$. Hence find the solution $u(x,t)$ first at $t = 0.01$ and then tabulate the results for $t = 0.05,0.10,...,0.25$ for the cases $x = 0.25,0.50,0.75$.

4. Find the solution $u(x,t)$ in Exercise 1 for the special case $f(x) = 4x(1-x)$ and $v = 0.01$. Tabulate this solution $u(x,t)$ for $t = 0.05,0.10,...,0.25$ for the cases $x = 0.25,0.50,0.75$.

5. Using the method of lines on Burgers' equation outlined in Section 5.3 (equations (5.29–5.31)), neglect the nonlinear term and use the initial condition

$$u(x,0) = \sin \pi x$$

to obtain the general solution

$$u = \alpha e^{\sqrt{B}x} + \beta e^{-\sqrt{B}x} + \left\{B/(B+\pi^2)\right\}\sin \pi x$$

where $B = 1/vk$ and α, β are constants to be determined.

Apply the boundary conditions

$$u(0,t) = u(1,t) = 0, \qquad t > 0,$$

and show that, to second order, the solution at the first time step is

$$u_1 = \frac{B}{B+\pi^2} \sin \pi x - \frac{\pi}{2(4\pi^2 + B)} \sin 2\pi x .$$

Hence find the first and second order approximations of u_1, at the first time step at $x = 0.25, 0.50, 0.75$ for the case $v = 1$ and $k = \delta t = 0.01$. Compare with the analytical results obtained in Exercise 3.

6. Check that the condition $0 < r \le 1/2$ is satisfied for the explicit FD scheme given in equation (5.18) where $u_{i,j} = u(ih, jk)$, $r = k/h^2$ for the case $v = 1$, $h = 0.25$, $k = 0.01$.

Use this explicit scheme to solve Burgers' equation at $x = 0.25, 0.50, 0.75$ for $t = 0.01, 0.02, \ldots, 0.25$. Check with the results from Exercises 3 and 5 at $t = 0.01$.

7. Use the implicit FD scheme in equation (5.19) to solve Burgers' equation taking $v = 1$, $h = 0.25$ and $k = 0.01$. Again check with the results from Exercises 3, 5 and 6 at $t = 0.01$.

Tabulate the solution $u(x,t)$ at $t = 0.05, 0.10, \ldots, 0.25$ and $x = 0.25, 0.50, 0.75$

8. Consider the solution of Burgers' equation using the method of lines described in Section 5.3. Write a computer program which uses a matrix inversion package to invert the matrix C in equation (5.40). Hence obtain results at $t = 0.01, 0.02, \ldots, 0.20$ for the case $v = 1$ using order $N = 4$.

Now build into the program the energy criterion of multiplying Burgers' equation throughout by u and integrating which was suggested by Cole [1] and use this to monitor the results.

Suggest how the results could be further improved.

9. Repeat Exercise 8 for the case $v = 0.1$, $N = 4$. Comment on the accuracy of the results.

10. Consider the solution of Burgers' equation under the boundary conditions

$$u(0,t) = u(1,t) = 0, \qquad t > 0.$$

(a) Use the FE method with fixed nodes ($n = 2$ case) with cubic shape functions for the cases:
  (i) $u(x,0) = \sin \pi x$ with $v = 1$;
  (ii) $u(x,0) = 4x(1 - x)$ with $v = 0.01$.
Obtain results at $t = 0.01, 0.02, \ldots, 0.25$ and compare with results previously obtained by other methods.

(b) Repeat (a) for the cases $n = 4$ and $n = 20$ and note the improvement in accuracy.

Bibliography

1. Cole, J.D., "On a quasi-linear parabolic equation occuring in Aerodynamics", Q. Appl. Math., 1951, **9**, 225.
2. Caldwell, J. and Wanless, P., "A Fourier Series approach to Burgers' equation", J.Phys.A: Math. Gen., 1981, **14**, 1029–1037.
3. Sincovec, R.F. and Madsen, N.K., "Software for Nonlinear Partial Differential Equations", ACM Transactions on Mathematical Software, 1975, **1**, 232.
4. Caldwell, J., "Solution of Burgers' equation by Fourier Transform methods", in *Wavelets, Fractals, and Fourier Transforms* (Eds: M. Farge, J.C.R. Hunt and J.C. Vassilicos), pp 309–316, Clarendon Press, Oxford, 1993.
5. Caldwell, J., Wanless, P. and Cook, A.E., "A Finite Element approach to Burgers' equation", Appl. Math. Modelling, 1981, **5**, 189–193.
6. Caldwell, J. and Smith, P., "Solution of Burgers' equation with a large Reynolds number", Appl. Math. Modelling, 1982, **6**, 381–385.
7. Caldwell, J., Wanless, P. and Cook, A.E., "Solution of Burgers' equation for large Reynolds number using Finite Elements with moving nodes", Appl. Math. Modelling, 1987, **11**, 211–214.
8. Caldwell, J., "Application of cubic splines to the nonlinear Burgers' equation", in *Numerical Methods for Nonlinear Problems*, vol. 3 (Eds: C. Taylor et al.), Pineridge Press, Swansea, 1986, 253–261.
9. Caldwell, J., Wanless, P. and Burrows, B.L., "A practical application of variational-iterative schemes", J. Phys. D: Appl. Phys., 1980, **13**, 177–179.

10. Saunders, R., Caldwell, J. and Wanless, P. "A variational-iterative scheme applied to Burgers' equation", IMA J. Num. Anal., 1984, **4**, 349–362.

11. Wang, P. and Kahawita, R., "Numerical integration of Partial Differential Equations using cubic splines", Int.J. Computer Math., 1983, **13**, 271–286.

12. Kato, T., "On Some Approximate Methods concerning the operators $T*T$ ", Math. Annln., 1953, **126**, p.253.

13. Walpole, L.J., "New Extremum Principles for Linear Problems", J. Inst. Math. Appl., 1974, **14**, p.113.

14. Arthurs, A.M., *Complementary Variational Principles*, Clarendon Press, Oxford, 1970.

15. Burrows, B.L. and Perks, A.J., "Complementary Variational Principles and Variational-Iterative Principles", J. Phys.A: Math. Gen., 1981, **14**, p.797.

Part II

CASE STUDIES

Case Study A

THE DESIGN OF BIOTREATMENT SYSTEMS

A.1 Introduction

Polluted water is discharged from a steelworks and the management have been ordered by the authorities to install an effluent system to 'clean up' the pollution before the water enters the local river. In particular, this system must ensure that the concentration of pollution is less than $5 \cdot 10^{-4}/m^3$ at all times. Now this could be a problem since the plant discharge concentration is about $10^{-2}/m^3$.

Biological treatment systems are generally considered to be the most reliable at dealing with steelworks' pollutant and one company has submitted a design utilizing a single well-mixed tank with a volume of 10,000 m^3 coping with a flow rate through the system of 10 m^3/hr. Some concern was expressed by the Steelworks Management at the tank size recommended by the biological treatment company. This concern generated two questions.

i) Will the proposed scheme always work?
ii) Could a more reliable system using two much smaller tanks be designed?

The purpose of this project is to provide some answers to these questions and to made a recommendation on the system to be used and its dimensions.

A.2 The Single Tank System

This straight forward system (Figure 2.7) consists of a well-mixed tank containing a volume V of water which has uniform concentrations of pollutant, $c(t)$, and organisms, $B(t)$. Polluted water, with a concentration of c^* flows into the tank at rate Q m^3/hr. and clean water flows out at the same rate.

Some basic micro biological investigations have revealed the following data on the growth and digestive characteristics of the organisms. The rate at which organisms multiply is proportional to both the organism and pollutant concentration. The birth rate constant is given by $R_2 = 1.2625$/hr. The death rate of organisms is proportional only to its concentration. The death rate constant is given by $D = 10^{-5}$/hr. The rate of pollutant digestion by the organisms is assumed proportional to both the organism and pollutant concentration and the relevant rate constant, $R_1 = 0.1$/hr.

A pair of ordinary differential equations may be constructed to describe the variation in pollutant and organism concentrations. These have been derived in Chapter 2 and are given by (2.23) and (2.24).

Now although there is a great deal of interest in devising efficient ways to start-up, etc., initially at least the basic question may be taken to mean that once a desirable operating condition has been reached will the proposed volume (i.e. 10,000 m^3) be adequate? If it is, how well will the system cope with changes in the incoming pollutant concentrations, c^*? The first of these questions may be answered by looking at the steady state solutions of equations (2.23) and (2.24) (i.e. when $dc/dt = 0$ and $dB/dt = 0$).

The equations give two solutions

a) $c = c^*$ and $B = 0$

b) $c = \left(D + \dfrac{Q}{V}\right)\bigg/ R_2$ and $B = \dfrac{Q}{VR_1}\left(c^* - c\right)\big/ c$

The first solution obviously corresponds to a "washout" situation and is undesirable. Using the available data, the second solution implies

$$c = \left(10^{-5} + \frac{10}{10^4}\right)\bigg/ 1.2625 \approx 10^{-3}$$

This means that the basic size of the tank is too small and will never reduce the pollutant to below the specified concentration of $5 \cdot 10^{-4}$. The tank volume required to do this at steady state would be

$$V = \frac{Q}{R_2 c - D} = \frac{10}{1.2625 \times 5 \times 10^{-4} - 10^{-5}} = 16097 \, \text{m}^3 \, .$$

Now we have to examine the stability of our system. It is possible to analyze the stability by computing the real parts of the eigenvalues of the Jacobian matrix evaluated at equilibrium. It is easy to compute the Jacobian of the system to be

$$J = \begin{bmatrix} -\dfrac{Q}{V} - R_1 B & -R_1 c \\ BR_2 & R_2 c - D - \dfrac{Q}{V} \end{bmatrix} .$$

For the equilibrium point ($c = (D + Q/V)/R_2$ and $B = Qc(c^* - c)/VR_1$), J reduces to

$$\begin{bmatrix} -\dfrac{Qc^* R_2}{DV + D} & -\dfrac{R_1(DV + Q)}{R_2 V} \\ -\dfrac{Q(c^* R_2 V + DV + Q)R_2}{VR_1(DV + Q)} & 0 \end{bmatrix} .$$

Using the available data, the eigenvalues are -0.0006318 and -0.0117932. It is clear that both eigenvalues have negative real part. Our system is considered to be stable at this equilibrium point. The single tank system was simulated using MATLAB program and the behaviour for the cases where c^* equals 0.01 with corresponding equilibrium concentration of organisms, B, is illustrated in Figure A.1. From this result, it is clear that although eventually the pollutant level falls below the legal limit there is a substantial time period where c is greater than $5 \cdot 10^{-4}$ (i.e. up to 3000 hours).

```
function yp = tank(t,y)
    D = 10^(-5); R1 = 0.1; R2 = 1.2625; Q = 10; V = 16097;
    c_input = 0.01;
    yp(1) = Q*(c_input-y(1))/V-R1*y(2)*y(1);
    yp(2) = y(2)*(R2*y(1)-D-Q/V);
end
```

```
function y = tank1
    t0 = 0; tfinal = 6000;
    y0 = [0 0.1180375];
    [t,y] = ode45('tank',t0,tfinal,y0,1.e-10);
    plot(t,y(:,1),[t0 tfinal],[5*10^(-4) 5*10^(-4)]);
    xlabel('time'); ylabel('c*');
end
```

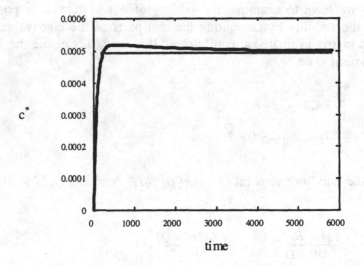

Figure A.1 $c^* = 0.01$ and $B = 0.1180375$ case.

In theory, a large tank should help to damp out such excursions. However, the tank size already exceeds the originally estimated size. We would like to keep the tank size unchanged. Different initial values of the concentration of organisms, $B(0)$, will also affect the behaviour of the system. A simple searching program for minimum initial concentration required to avoid the pollutant concentration exceeding the legal limit at all times can be established by MATLAB and the results are illustrated in Figure A.2. From these results we obtain some ideas on how the initial concentration of organisms should be set up. When $B(0)$ is equal to 0.12435, the pollutant level will meet the legal criteria at all times with the incoming concentration equal to 0.01.

```
function y = tank1
    t0 = 0; tfinal = 15000;
    Flag = 0;
    y0 = [0 0.1180375];
    while Flag == 0
```

```
            [t,y] = ode45('tank',t0,tfinal,y0,1.e-10);
            if (y(:,1)-5*10^(-4)*ones(size(y,1),1))<=0
                Flag = 1;
            else
                y0 = y0 + [0 0.00001]';
            end
        end
        plot(t,y(:,1),[t0 tfinal],[5*10^(-4) 5*10^(-4)]));
        xlabel('time'); ylabel('c*');
end
```

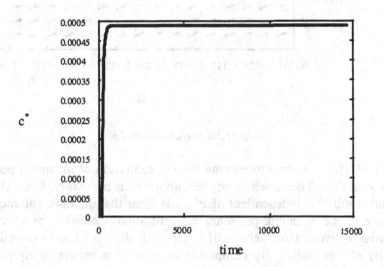

Figure A.2 $c^* = 0.01$ and minimum $B(0)$ required $= 0.12435$.

It can be easily seen from a directional field plot (see Figure A.3) that if you start at a point X (i.e. $c = 0$ and $B = 0.1180375$), pollutant level, c, will tend to increase and will eventually pass the line $c = 0.0005$ (i.e. the legal limit) and move back to the steady point. However, if you start at a point Y (i.e. $c = 0$ and $B = 0.12435$), c will increase directly to the steady point and never exceed 0.0005.

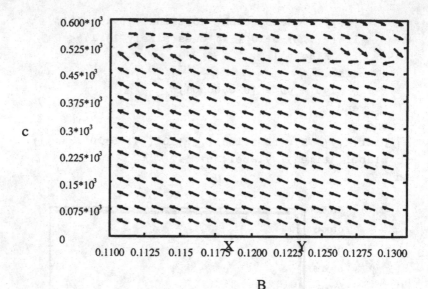

Figure A.3 Directional field plot

In real life, it is usual to assume the concentration of incoming pollutant varies with time. Then a sensitivity test about c^* is necessary. Since the tank volume required is independent of c^*, it is clear that the tank volume is not sensitive to the incoming pollutant concentration. Consider the case when c^* varies between 0.005 and 0.015. First of all, we have to examine the stability of our system. By computing the Jacobian matrix at equilibrium, the eigenvalues of the system are

$$0.621c^* \pm 0.967 \cdot 10^{-8} \sqrt{0.413 \cdot 10^{16} c^{*2} - 0.839 \cdot 10^{13} c * + 0.420 \cdot 10^{10}}$$

It can be easily shown that both eigenvalues have negative real part. Our system is considered to be stable at this equilibrium point for all time.

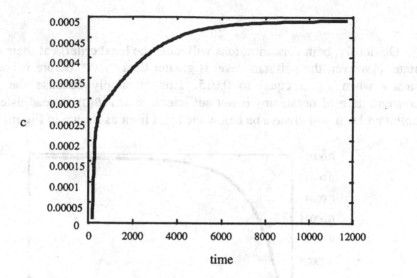

Figure A.4 c* =0.005, V = 16097 case

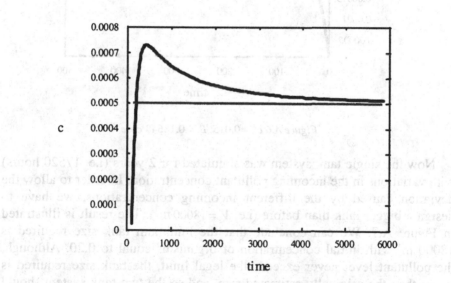

Figure A.5 c* =0.015, V = 16097 case

Next, we can simulate the system behaviour for the extreme cases where c^* equals 0.005 and 0.015. The results are illustrated in Figures A.4 and A.5, respectively.

Obviously, both concentrations will meet the legal criteria at their steady state. However, the pollutant level is greater than $5 \cdot 10^{-4}$ before it becomes steady when c^* is equal to 0.015. This is simply because the initial concentration of organisms is not sufficient. When $B(0)$ is readjusted, the pollutant level will always be below the legal limit as shown in Figure A.6.

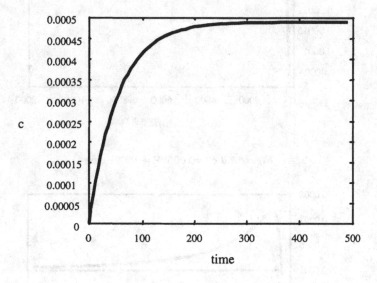

Figure A.6 c^* =0.015, B = 0.18648 case

Now the single tank system was simulated for 2 years (i.e. 17520 hours) with variations in the incoming pollutant concentration. In order to allow the deviation caused by the different incoming concentrations, we have to design a bigger tank than before (i.e. $V = 18000 \, \text{m}^3$). The result is illustrated in Figure A.7. We can conclude that the minimum tank size required is 18000 m^3 with initial concentration of organisms equal to 0.20. Although, the pollutant level never exceeds the legal limit, the tank size required is larger than the originally estimated size, and so the two-tank system should be considered.

```
function f1 = f1(a,b);
    R1 = 0.1; Q = 10; V = 18000;
    C_input = 0.005 + rand*0.01;
    f1 = Q*(c_input-a)/V-R1*a*b;
end
```

```
function f2 = f2(a,b);
    D = 10^(-5); R2 = 1.2625; Q = 10; V = 18000;
    f2 = b*(R2*a-D-Q/V);
end



function M = rk4(N,T)
    x = 0; y = .20; t = 0;
    M = zeros(N,3);
    h = (T-t)/N;
    for n = 1:N
        M(n,1) = x; M(n,2) = y; M(n,3) = t;
        r1 = f1(x,y);
        s1 = f2(x,y);
        r2 = f1(x+(h/2)*r1,y+(h/2)*s1);
        s2 = f2(x+(h/2)*r1,y+(h/2)*s1);
        r3 = f1(x+(h/2)*r2,y+(h/2)*s2);
        s3 = f2(x+(h/2)*r2,y+(h/2)*s2);
        r4 = f1(x+h*r3,y+h*r3);
        s4 = f2(x+h*r3,y+h*r3);
        x = x+(h/6)*(r1+2*r2+2*r3+r4);
        y = y+(h/6)*(s1+2*s2+2*s3+s4);
        t = t+h;
    end
    plot(M(:,3),M(:,1));
    axis([0 T 0 5*10^(-4)]);
    xlabel('time'); ylabel('c');
end
```

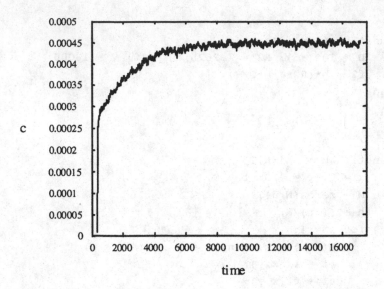

Figure A.7 Simulation for 17520 hours with volume = 18000.

A.3 The Two Tank System

This system is illustrated in Figure A.8. The tanks are assumed to be of volume V_1 and V_2 with the outflow from tank one forming the inflow to tank two. Furthermore, there is no reason to suppose that the growth and digestive behaviour of the organisms are any different to that for the single tank system.

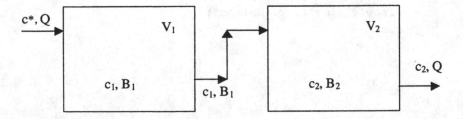

Figure A.8 The two tank system

The behaviour of the organism and pollutant concentration in each of the tanks may be described by a system of four ordinary differential equations. The mathematical description of the first tank is formally the same as for the single tank system. Hence,

$$\frac{dc_1}{dt} = \frac{Q}{V_1}\left(c^* - c_1\right) - R_1 B_1 c_1 \qquad (A.1)$$

$$\frac{dB_1}{dt} = B_1\left(R_2 c_1 - D - \frac{Q}{V_1}\right) \qquad (A.2)$$

The change in pollutant concentration in tank two is given by

$$\begin{bmatrix} \text{Change in pollutant} \\ \text{level in tank two} \end{bmatrix} = \begin{bmatrix} \text{pollutant arriving} \\ \text{from inflow} \end{bmatrix} - \begin{bmatrix} \text{pollutant leaving in} \\ \text{outflow} \end{bmatrix} - \begin{bmatrix} \text{pollutant digested} \\ \text{by organisms} \end{bmatrix}$$

which may be written mathematically as

$$\frac{dc_2}{dt} = \frac{Q}{V_2}\left(c_1 - c_2\right) - R_1 B_2 c_2 \qquad (A.3)$$

The corresponding variation in the organism population is given by

$$\begin{bmatrix} \text{Change in organism} \\ \text{population in tank two} \end{bmatrix} = \begin{bmatrix} \text{organisms created} \\ \text{from reproduction} \end{bmatrix} - \begin{bmatrix} \text{organisms lost} \\ \text{by death} \end{bmatrix}$$
$$+ \begin{bmatrix} \text{organisms arriving in} \\ \text{inflow from tank one} \end{bmatrix} - \begin{bmatrix} \text{organisms lost} \\ \text{in the outflow} \end{bmatrix}$$

which may be written mathematically as

$$\frac{dB_2}{dt} = B_2\left(R_2 c_2 - D\right) + \frac{Q}{V_2}\left(B_1 - B_2\right) \qquad (A.4)$$

Equations (A.1)-(A.4) constitute the model for the two tank system. The problem now comes down to using these equations to evaluate V_1 and V_2 in such a way that $c_2 < 5 \cdot 10^{-4}$ under all possible variations in operating conditions.

As for the single tank, it is useful to look at the steady state situations for some basic design criteria. Equations (A.1) and (A.2) yield the same solutions for B_1 and C_1 as in the single tank case, i.e.

a) $c_1 = c^*$ $B_1 = 0$

$$(A.5)$$

b) $\quad c_1 = \left(D + \dfrac{Q}{V}\right)\Big/ R_2 \qquad\qquad B_1 = \dfrac{Q}{V_1 R_1}\left(c^* - c_1\right)\Big/ c_1$

Equation A.5(b) places a constraint on the minimum size of tank one. Obviously $c_1 \le c^*$, and hence A.5(b) implies

$$c^* \ge \left(D + \dfrac{Q}{V_1}\right)\Big/ R_2$$

i.e. $\quad V_1 = Q/\left(R_2 c^* - D\right)$ \hfill (A.6)

If $c^* = 0.01$ then $V_1 \ge 792.7\,\mathrm{m}^3$. If V_1 is less than either of these values (for corresponding stable conditions) then $c_1 = c^*$ and $B_1 = 0$, and the pollutant passes through untreated. It is useful to consider why this is so. From equation (A.2) it is clear that if $-Q/V_1$ is too large it will dominate over the organism population growth rate (i.e. $R_2 c_1$) and all the organisms will be washed out in the outflow. It is this effect that gives rise to the minimum possible volume for the first tank.

Assuming the pollutant is treated in the first tank, the solution of the second pair of equations yields

$$B_2 = \dfrac{Q}{V_2}\,\dfrac{\left(c_1 - c_2\right)}{R_1 c_2}$$ \hfill (A.7)

if it is assumed that $c = 5 \cdot 10^{-4}$ (i.e. the prescribed upper legal limit on the pollutant level discharged to the river). Then equation (A.7) (with only V_1 unknown) may be substituted into the steady state form of equation (A.4), i.e.

$$\dfrac{\left(c_1 - c_2\right)}{R_1 c_2}(R_2 c_2 - D) + \left(B_1 - \dfrac{Q}{V_2}\,\dfrac{\left(c_1 - c_2\right)}{R_1 c_2}\right) = 0$$

This equation may be manipulated to give

$$V_2 = \dfrac{Q(c_1 - c_2)}{\left(c_1 - c_2\right)\left(R_2 c_2 - D\right) + \left(B_1 R_1 c_2\right)}$$ \hfill (A.8)

where, for a given value of V_1, both B_1 and c_1 may be calculated from equation (A.5) and all the other parameters are known.

A small program was written to calculate V_2 over a range of V_1 values for steady state conditions at $c^* = 0.01$. A listing of the program and a graph of the results are shown in Figure A.9. These results are also prescribed in Table A.1.

```
function volume
    D=10^(-5); R1=0.1; R2=1.2625; Q=10; limit=5*10^(-4);
    c_input = 0.01; c2 = limit;
    for i=1:970
        V1(i)=i*20+800;
        c1(i)=(D+Q/V1(i))/R2;
        B1(i)=Q*(c_input-c1(i))/(V1(i)*R1*c1(i));
        V2(i)=Q*(c1(i)-c2)/((c1(i)-c2)*(R2*c2-D)+B1(i)*R1*c2);
        B2(i)=Q*(c1(i)-c2)/(V2(i)*R1*c2);
        Total(a)=V1(i)+V2(i);
    end
    plot(V1,Total,[V1(1),V1(970)],[16097,16097]);
    xlabel('Volume 1'); ylabel('Total Volume');
end
```

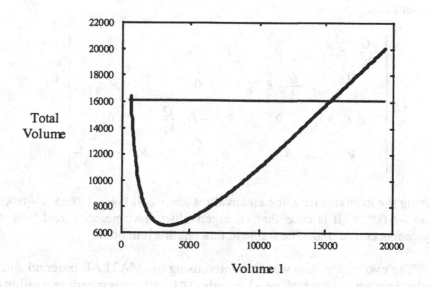

Figure A.9 $c^* = 0.01$ case

Volume 1	Volume 2	Total Volume	B1	B2	c1
3600	2865	6465	0.098019	0.119243	0.002208
3610	2855	6465	0.098095	0.119243	0.002202
3620	2845	6465	0.098170	0.119243	0.002196
3630	2834	6464	0.098245	0.119243	0.002190
3640	2824	6464	0.098320	0.119243	0.002184
3650	2814	6464	0.098394	0.119243	0.002178
3660	2804	6464	0.098467	0.119243	0.002172
3670	2795	6465	0.098540	0.119243	0.002166
3680	2785	6465	0.098613	0.119243	0.002160
3690	2775	6465	0.098686	0.119242	0.002154
3700	2765	6465	0.098758	0.119242	0.002149

Figure A.9 Results for V_2 over a range of V_1 with $c^* = 0.01$

From these result it is clear that smaller combined tank volumes than for the single tank system may be yielded. The minimum combined tank volumes will be obtained when $V_1 = 3650$ and $V_2 = 2814$, with total volume $= 6464 < 16097$ m^3.

Now we have to examine the stability of our system. The Jacobian of the system is

$$
J = \begin{bmatrix}
-\dfrac{Q}{V_1} - R_1 B_1 & 0 & -R_1 c_1 & 0 \\
\dfrac{Q}{V_2} & -\dfrac{Q}{V_2} - R_1 B_2 & 0 & -R_1 c_2 \\
B_1 R_2 & 0 & R_2 c_1 - D - \dfrac{Q}{V_1} & 0 \\
0 & B_2 R_2 & \dfrac{Q}{V_2} & R_2 c_2 - D - \dfrac{Q}{V_2}
\end{bmatrix}
$$

Using the available data, the eigenvalues are -0.01484, -0.00356, -0.00983 and -0.00275. It is clear that all eigenvalues have negative real part. Our system is considered to be stable at this equilibrium point.

The two tank system was simulated using the MATLAB program and the behaviour for the case where c^* equals 0.01 with corresponding equilibrium concentration of organisms, B, is illustrated in Figure A.10.

```
function yp = tank2(t,y)
```

```
      D=10^(-5); R1=0.1; R2=1.2625; Q=10; V1=3650; V2=2814;
      c_input=0.01;
      yp(1) = Q*(c_input-y(1))/V1-R1*y(2)*y(1);
      yp(2) = y(2)*(R2*y(1)-D-Q/V1);
      yp(3) = Q*(y(1)-y(3))/V2-R1*y(4)*y(3);
      yp(4) = y(4)*(R2*y(3)-D)+Q*(y(2)-y(4))/V2;
   end


   function y = tank
      t0 = 0; tfinal = 3000;
      y0 = [0 0.09839 0 0.11924];
      [t,y] = ode45('tank2',t0,tfinal,y0,1.e-10);
      plot(t,y(:,3),[t0 tfinal],[5*10^(-4) 5*10^(-4)]);
      xlabel('time'); ylabel(c2);
   end
```

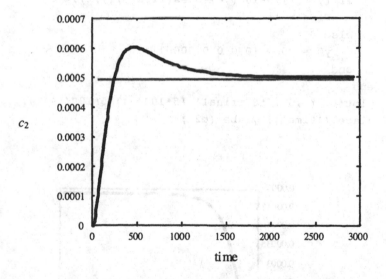

Figure A.10 $B1(0) = 0.09839$, $B2(0) = 0.11924$, $V1 = 3650$, $V2 = 2814$ *case*

Again, from this result it is clear that although eventually the pollutant level falls below the legal limit there is a substantial time period when c is greater than 5×10^{-4} (i.e. up to 2500 hours). We can search for the best initial value of the concentration of organisms in both tanks, $B1(0)$ and $B2(0)$, by either fixing $B1(0)$ and varying $B2(0)$ or fixing $B2(0)$ and varying $B1(0)$ or changing both $B1(0)$ and $B2(0)$ at the same time. The results are illustrated in Figure A.11. When we fixed $B2(0)$ as 0.11924 and changed $B1(0)$ from

0.09839 to 0.12690, the pollutant level will meet the legal criteria at all time.

```
function y = tank
    t0 = 0; tfinal = 1500;
    Flag = 0;
    y0 = [0 0.09839 0 0.11924];
    while Flag == 0
        [t,y] = ode45('tank2',t0,tfinal,y0,1.e-10);
        if (y(:,1)-5*10^(-4)*ones(size(y,1),1))<=0
            Flag = 1;
        else
            y0 = y0 + [0 0.00001 0 0]'
        end
        [t,y] = ode45('tank2',t0,tfinal,y0,1.e-10);
        if (y(:,3)-5*10^(-4)*ones(size(y,1),1))<=0
            Flag = 1;
        else
            y0 = y0 + [0 0 0 0.00001]'
        end
    end
    plot(t,y(:,3),[t0 tfinal],[5*10^(-4) 5*10^(-4)]);
    xlabel('time'); ylabel(c2);
end
```

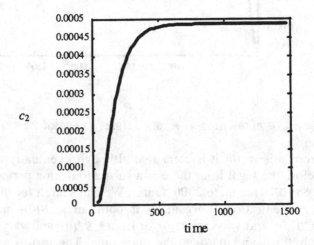

c_2

Figure A.11 B1(0)=0.12690 , B2(0)=0.11924 *case*

If the dischange pollutant level could be kept constant then a two tank system with $V_1 = 3650$ m^3 and $V_2 = 2814$ m^3 would be a better choice than a single tank system with volume equal to 16097 m^3. If variation of c^* is considered, a sensitivity test about c^* is necessary. From equation (A.6), it is clear that the value of c^* has a critical effect on the tank sizes required to treat the pollutant satisfactorily. A list of results for the two extreme cases, $c^* = 0.005$ and $c^* = 0.015$, are shown in Figure A.12 and A.13. If $c^* = 0.005$ then $V_1 \geq 1586.7$ m^3 whilst if $c^* = 0.015$ then $V_1 \geq 528.3$ m^3. If $V1$ is less than either of these values then $c1 = c^*$ and $B1 = 0$, and the pollutant passes through untreated. This means that all designs must be based upon a c^* value of 0.005 so that the worst cases can always be dealt with. This obviously means using the results shown in Figure A.12.

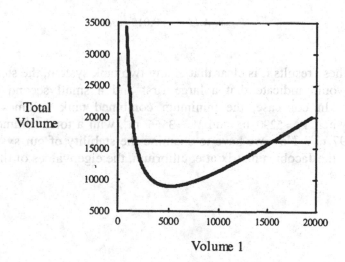

Figure A.12 $c^* = 0.005$ case

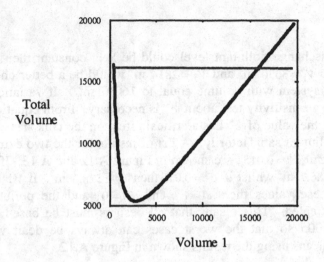

Figure A.13 c* = 0.015 case

From these results it is clear that in any two tank system, the steady state analysis would indicate that a large first and a small second tank are desirable. In our case, the minimum combined tank volumes will be obtained when $V_1 = 5290$ m^3 and $V_2 = 3566$ m^3, with a total volume $= 8856$ m$^3 < 16097$ m^3. Now we have to examine the stability of our system. By computing the Jacobian matrix at equilibrium, the eigenvalues of the system are

$$eig1 = -0.830*10^{-6}\%1\left(-41556.11\frac{1}{\%1} - 374.166\frac{1}{\%1}\right.$$

$$+602278.576\frac{\sqrt{0.00187\%1^2 + 0.123*10^{-5} - 0.00161c^*}}{\%1^2}$$

$$eig2 = -0.830*10^{-6}\%1\left(-41556.11\frac{1}{\%1} - 374.166\frac{1}{\%1}\right.$$

$$-602278.576\frac{\sqrt{0.00187\%1^2 + 0.123*10^{-5} - 0.00161c^*}}{\%1^2}$$

$$\%1 = 0.0151 - 29.089c^*$$

$$eig3 = -0.629c^*$$

$$+\frac{1}{2}\sqrt{\frac{16000}{2147189078241}\left(\frac{20073}{2000} - 4608.125c^*\right)^2 + \frac{15073}{666125000} - 0.00694c^*}$$

$$eig4 = -0.629c^*$$

$$-\frac{1}{2}\sqrt{\frac{16000}{2147189078241}\left(\frac{20073}{2000} - 4608.125c^*\right)^2 + \frac{15073}{666125000} - 0.00694c^*}$$

As before it can easily be shown that all eigenvalues have negative real part. Our system is considered to be stable at all time.

Next, we can simulate the system behavior for the extreme cases where c^* equals 0.005 and 0.015. The results are illustrated in Figure A.14 and A.15.

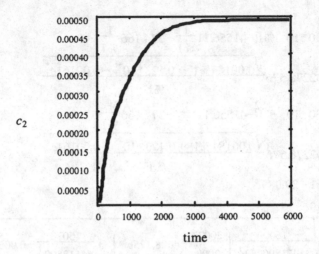

Figure A.14 c* = 0.005, V_1 = 5290, V_2 = 3566 case

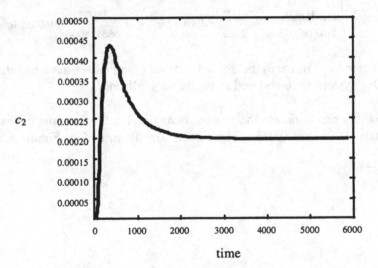

Figure A.15 c* = 0.015, V_1 = 5290, V_2 = 3566 case

From these results, clearly the pollutant level will always fall below the legal limit all the time. Various c^* from 0.01 to 0.015 were tried. The pollutant level still always meets the legal criteria. There is no need to readjust the initial concentration of organism at this time. Now the two tank system was simulated for 2 years (i.e. 17520 hours) with variations in the incoming pollutant concentration. The results are illustrated in Figure A.16.

```
function f1 = f1(c1,B1,c2,B2);
   R1 = 0.1; Q = 10; V1 = 5290;
   c_input = 0.005 + rand*0.01;
   f1 = Q*(c_input-c1)/V1-R1*B1*c1;
end


function f2 = f2(c1,B1,c2,B2);
   D = 10^(-5); R2 = 1.2625; Q = 10; V1 = 5290;
   f2 = B1*(R2*c1-D-Q/V1);
end


function f3 = f3(c1,B1,c2,B2);
   R1 = 0.1; Q = 10; V2 = 3566;
   f3 = Q*(c1-c2)/V2-R1*B2*c2;
end

function f4 = f4(c1,B1,c2,B2);
   D = 10^(-5); R2 = 1.2625; Q = 10; V2 = 3566;
   f4 = B2*(R2*c2-D)+Q*(B1-B2)/V2;
end


function M = rk4(N,T);
   c1 = 0; B1 = 0.1269; c2 = 0; B2 = 0.11924; t = 0;
   M = zeros(N,5);
   h = (T-t)/N;
   for n = 1:N
      M(n,1) = c1; M(n,2) = B1; M(n,3) = c2; M(n,4) = B2;
      M(n,5) = t;
      r1 = f1(c1,B1,c2,B2);
      s1 = f2(c1,B1,c2,B2);
      u1 = f3(c1,B1,c2,B2);
      v1 = f4(c1,B1,c2,B2);
      r2 = f1(c1+h*r1/2,B1+h*s1/2,c2+h*u1/2,B2+h*v1/2);
      s2 = f2(c1+h*r1/2,B1+h*s1/2,c2+h*u1/2,B2+h*v1/2);
      u2 = f3(c1+h*r1/2,B1+h*s1/2,c2+h*u1/2,B2+h*v1/2);
      v2 = f4(c1+h*r1/2,B1+h*s1/2,c2+h*u1/2,B2+h*v1/2);
      r3 = f1(c1+h*r2/2,B1+h*s2/2,c2+h*u2/2,B2+h*v2/2);
      s3 = f2(c1+h*r2/2,B1+h*s2/2,c2+h*u2/2,B2+h*v2/2);
      u3 = f3(c1+h*r2/2,B1+h*s2/2,c2+h*u2/2,B2+h*v2/2);
      v3 = f4(c1+h*r2/2,B1+h*s2/2,c2+h*u2/2,B2+h*v2/2);
      r4 = f1(c1+h*r3,B1+h*s3,c2+h*u3,B2+h*v3);
      s4 = f2(c1+h*r3,B1+h*s3,c2+h*u3,B2+h*v3);
      u4 = f3(c1+h*r3,B1+h*s3,c2+h*u3,B2+h*v3);
```

```
     v4 = f4(c1+h*r3,B1+h*s3,c2+h*u3,B2+h*v3);
     c1 = c1+(h/6)*(r1+2*r2+2*r3+r4);
     B1 = B1+(h/6)*(s1+2*s2+2*s3+s4);
     c2 = c2+(h/6)*(u1+2*u2+2*u3+u4);
     B2 = B2+(h/6)*(v1+2*v2+2*v3+v4);
     t = t+h;
   end
   plot(M(:,5),M(:,3),axis([0 T 0 5*10^(-4)]),...
   xlabel('time'), ylabel('t')
end
```

From the results, the pollutant becomes steady at below the level we expected (i.e. $5 \cdot 10^{-4} / m^3$). It is simply because the tanks are too large for the cases when c^* is greater than 0.005. However, we would not like to reduce the tank sizes in order that the worst cases can always be dealt with. (i.e. the incoming pollutant stays constant at the level of 0.005 for a very long time period).

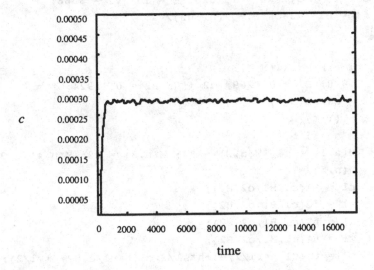

Figure A.16 Simulation for 17520 *hours with* $V_1 = 5290$, $V_2 = 3566$

A.4 Conclusion

The biological treatment design recommending a single tank of volume 10000 m^3 is totally inadequate to process the pollutant concentration discharged from the steelworks. If the discharge pollutant level could be kept constant then a tank size of about 16000 m^3 would be required. Since the pollutant level will vary, then to avoid exceeding the legal limit for long time periods a bigger tank would be required. The single tank system is not completely practical.

The equation constituting the model for the two tank system shows that to avoid washout from tank one its volume must exceed 1600 m³. In fact, the avoidance of washout places a serious constraint on the tank sizes that can be utilized. This factor concentrates attention on the design aspects to the situation when the incoming pollution level is low (i.e. 0.005), for it is here that the tank volumes need to be largest. The system then has to be coping with variation from 0.005 to 0.015 in the pollutant level.

The model shows that it is possible to always keep the exit pollutant concentration less than $5 \cdot 10^{-4}/m^3$ provided $V1 = 5290$, $V2 = 3566$, $B1(0) = 0.12690$ and $B2(0) = 0.11924$. Obviously, it is much more effective at processing the pollutant level than in the single tank system.

Finally, it is worth noting that if the Steelworks pollutant level could be maintained near 0.015 then a much smaller system would be required to process it effectively.

A.5 Problems

1.   In this Case Study on the design of biotreatment systems consider the single tank system discussed in Section A.2. In particular, consider the case when $c*$ varies between 0.005 and 0.015.
   (a)   Examine the stability of the system by computing the Jacobian matrix at equilibrium.
   (b)   Show that both eigenvalues have negative real part by plotting computer graphs for a range of values of $c*$.

1.   Consider the two tank system for the design of biotreatment systems as discussed in Section A.3.
   (a)   Confirm that the minimum conbined tank volume will be obtained when $V_1 = 5290m^3$ and $V_2 = 3566m^3$, with a total volume $V = 8856m^3$.
   (b)   Compute the Jacobian matrix at equilibrium and hence obtain the 4 eigenvalues of the system.
   (c)   Show that all 4 eigenvalues have negative real part by plotting computer graphs for a range of values of $c*$.

Bibliography

1.   Braun, M., *Differential Equations and Their Applications*, Springer-Verlag, 4th ed., 1993.
2.   Burghes, D.N., Huntley, I. and McDonald, J., *Applying Mathematics: A course in mathematical modelling*, Ellis Horwood Ltd., 1982.
3.   Cross, M. and Moscardini, A.O., *Learning the Art of Mathematical Modelling*, Ellis Horwood Limited, 1985.
4.   Farlow, S.J., *An Introduction to Differential Equations and Their Application*, McGraw-Hill, Inc., 1994.
5.   Griffiths, H.B. and Oldknow, A., *Mathematics of Models: Continuous and Discrete Dynamical Systems*, Ellis Horwood Ltd., 1993.
6.   Klamkin, M.S., *Mathematical Modelling: Classroom Notes in Applied Mathematics*, SIAM, 1987.
7.   Meerschaert, M.M., *Mathematical Modeling*, Academic Press, Inc., 1993.

Case Study B
DYNAMICS OF CONTAGIOUS DISEASE

Summary

This case study considers a deterministic model in the theory of contagious disease. The major assumption made is that all infected individuals remain in the population to spread the disease. For this particular case an exact solution can be obtained for the differential equation arising from the mathematical model. However, a more generally useful model can be devised by introducing certain numerical tools for simulating the solutions, namely, error control of difference methods and, in particular, the use of the Runge-Kutta-Fehlberg numerical method. This algorithm is found to give highly accurate results and important conclusions are drawn from this modelling and numerical study.

B.1 Introduction

This study involues the approximation of the solution of an initial-value problem of the general form

$$y' = f(t, y), \quad a \leq t \leq b, \quad y(a) = \alpha \tag{B.1}$$

which arises in the theory of contagious disease. In this case a differential equation can be used to predict the number of infections individuals in the population at any time provided appropriate simplifying assumptions are made.

The case study will examine a technique that can be used to control the error of a difference method for approximating the solution of an initial value problem in an efficient manner by the appropriate choice of mesh points.

This will involve determining how the estimation of the local truncation error of a difference method can be used to advantage in approximating the optimal step size to control the global error. A popular technique for error control presented by Fehlberg [1] called the Runge-Kutta-Fehlberg method will be considered. Essentially this technique consists of using a Runge-Kutta method with local truncation error of order 5 to estimate the local error in a Runge-Kutta method of order 4. One clear advantage of this method is that only 6 evaluations of the function f are required per step, whereas arbitrary Runge-Kutta methods of order 4 and 5 used together would require 10 evaluations of f per step.

An algorithm and computer program using the Runge-Kutta-Fehlberg method with error control will be devised to approximate the solution of the initial-value problem (B.1) with local truncation error with a given tolerance. This algorithm will be tested out and applied in approximating the solution of the differential equation model for contagious disease (see Bailey [2–3]).

B.2 Contagious Disease Problem

General introduction. The gap that exists between the ideal of complete well being for all peoples and the overall level of achievement in the world is massive. Medical workers and health authorities are putting huge efforts and resources into trying to rectify this situation. Within this environment, mathematicans can be of invaluable assistance, in particular, in the area of allocation of resources. Clearly, the ability to predict how a disease will develop and spread (if at all) will greatly benefit workers in this area.

Various mathematical and statistical investigations are used in medicine. Historically, they were begun by Graunt and Petty who studied the London Bills of Mortality in the 17th centry, and by the end of the 18th century Daniel Bernoulli used a mathematical method to evaluate the effectiveness of the variolation technique against smallpox.

It is important therefore to study a deterministic model of an epidemic – i.e., the sort of model that, given certain initial conditions, can predict how an epidemic will spread.

Formulation of a deterministic model. The simplest scenario for an epidemic is one in which the population is split into two parts, one is the infected (who are also **all** infectious) and the other is the part that is susceptible of being infected. There are no people who are immune and all the infected people are infectious and vice versa. The infection is spread by contact between members of the community in which there is no removal from circulation by death, recovery or isolation. Ultimately, therefore, all suspceptibles become infected.

Although these assumptions are too simplified for many practical cases they are applicable to the sort of situation where the disease is highly infectious but not sufficiently serious for cases to be withdrawn by death or isolation and no infective becomes clear of infection during the main part of the epidemic. This might be the case for some of the milder infections of the upper respiratory tract. (see Bailey [3]).

We can construct a deterministic model for this situation as discussed by Bailey [3]. Suppose we have a homogeneously mixing group of individuals of size $n + a$, and suppose that the epidemic is started off at time $t = 0$ by a individuals becoming infected, the remaining n individuals all being susceptiable, but as yet unifected.

Let $x(t)$ and $y(t)$ be continuous, where

$x(t)$ = number of susceptibles,

$y(t)$ = number of infectives.

Suppose also that the rate of occurrence of new infections is proportional to both the number of infectives and the number of susceptibles. Then we can write

$$\Delta y = \beta xy \Delta t$$

where

Δy = increase in number of infectives,
β = infective rate (or contact rate),
Δt = increase in time.

This reduces to the differential equation,

$$\frac{dy}{dt} = \beta x(t) y(t), \qquad y(0) = a, \qquad x(0) = n.$$ (B.2)

Clearly, $x(t) + y(t) = n + a$ for all t, and therefore,

$$\frac{dy}{dt} = \beta y(t)\{n + a - y(t)\}, \qquad y(0) = a.$$ (B.3)

B.3 Analytic Solution

The above differential equation takes the form of a Bernoulli equation which has a standard technique of solution. The following method makes use of this technique.

We have the differential equation,

$$\frac{dy}{dt} = \beta y [n + a - y], \qquad y(0) = a$$

$$\frac{dy}{dt} = -\beta y^2 + \beta [n + a] \cdot y$$

$$\therefore \qquad \frac{1}{y^2} \cdot \frac{dy}{dt} = -\beta + \beta [n + a] \frac{1}{y}$$

$$\therefore \qquad \frac{1}{y^2} \cdot \frac{dy}{dt} - \beta [n + a] \frac{1}{y} = -\beta.$$ (B.4)

Let $z = 1/y \quad \Rightarrow \quad z(0) = 1/a$

$$\frac{dz}{dt} = \frac{1}{y^2} \cdot \frac{dy}{dt} \qquad \Rightarrow \qquad \frac{1}{y^2} \cdot \frac{dy}{dt} = -\frac{dz}{dt}.$$

Substitute into (B.4) for z and dz/dt to obtain

$$-\frac{dz}{dt} - \beta [n + a] \cdot z = -\beta$$

i.e.

$$\frac{dz}{dt} + z\beta[n+a] = \beta.$$ (B.5)

We now need an integrating factor (IF) to solve this equation, namely

$$IF = e^{\int \beta(n+a)dt} = e^{\beta(n+a)t}.$$

Multiply both sides of (B.5) by IF to obtain

$$\left(\frac{dz}{dt} + z\beta[n+a]\right)e^{\beta(n+a)t} = \beta e^{\beta(n+a)t}.$$

Integating gives

$$ze^{\beta(n+a)t} = \beta \int e^{\beta(n+a)t} dt$$

$$= \frac{\beta e^{\beta(n+a)t}}{n+a} + C, \quad C = \text{constant of integration}$$

$$= \frac{e^{\beta(n+a)t}}{n+a} + C.$$

Now $z(0) = 1/a$,

$$1/a = 1/(n+a) + C \implies C = \frac{n}{a(n+a)}.$$

Therefore,

$$ze^{\beta(n+a)t} = \frac{e^{\beta(n+a)t}}{n+a} + \frac{n}{a(n+a)}.$$

Hence,

$$z = \frac{ae^{\beta(n+a)t} + n}{a(n+a)e^{\beta(n+a)t}}$$

$$= \frac{a + ne^{-\beta(n+a)t}}{a(n+a)}.$$

But we know that $y = 1/z$ and thus we have the following formula:

$$y(t) = \frac{a(n+a)}{a + ne^{-\beta(n+a)t}}.$$ (B.6)

This formula shows that the number of infectives at any time t depends only on the initial conditions at $t = 0$ and the infection rate β. This is what we would expect from a deterministic model.

B.4 Error Control of Difference Methods

Estimation of local truncation error. In approximating the solution $y(t)$ of the initial-value problem given by equation (B.1), an ideal difference equation method

$$w_{i+1} = w_i + h_i \phi(t_i, h_i, w_i), \quad i = 0, 1, \ldots, N-1,$$ (B.7)

would have the property that, whenever a tolerance $\xi > 0$ was given, the minimum number of mesh points would be used to ensure that the global error $|y(t_i) - w_i|$ did not exceed ξ for any $i = 0, 1, 2, \ldots, N$. To ensure a minimal number of mesh points and also control the global error of a difference method will mean that the points will not necessarily be equally spaced in the interval. We will therefore consider a technique which can be used to control the error of a difference method in an efficient manner by the appropriate choice of mesh points.

It can be shown (see Maron [4], Gerald and Wheatley [5] and Rice [6]) that the local truncation error of a method can be controlled by controlling the global error. In fact, the global error and the local error have the same rate of convergence. It would therefore be useful to have an estimate of the local error.

We will therefore consider two methods of approximating the solution of an initial value problem, the first method (A) having a local error of order n and the second method (B) a local error of order $n+1$.

Let w_i be the set of approximations generated by method A and $\overline{w}_i$ be the corresponding set generated by method B. The actual set of solutions will be $y(t_i)$ which we denote by y_i. Using h for the step increase at the i th stage we therefore have the following:

Method A

$$w_0 = \alpha$$
$$w_{i+1} = w_i + h\phi(t_i, h, w_i).$$

Hence

$$y_{i+1} = y_i + h\phi(t_i, h, w_i) + hT_{i+1}(h^n),$$

where T_{i+1} is the local error. Therefore

$$hT_{i+1} = y_{i+1} - y_i - h\phi(t_i, h, y_i).$$

Method B

$$\overline{w}_0 = \alpha$$
$$\overline{w}_{i+1} = \overline{w}_i + h\phi(t_i, h, \overline{w}_i).$$

Now, if $w_i \approx y_i \approx \overline{w}_i$, then

$$
\begin{aligned}
y_{i+1} - w_{i+1} &= y_{i+1} - w_i - h\phi(t_i, h, w_i) \\
&\approx y_{i+1} - y_i - h\phi(t_i, h, y_i) \\
&= hT_{i+1}.
\end{aligned}
$$

Therefore

$$
\begin{aligned}
hT_{i+1} &\approx y_{i+1} - w_{i+1} \\
&= (y_{i+1} - \overline{w}_{i+1}) + (\overline{w}_{i+1} - w_{i+1}) \\
&\approx \overline{T}_{i+1} + \overline{w}_{i+1} - w_{i+1}.
\end{aligned}
$$

But T_{i+1} is of order h^n while $\overline{T}_{i+1}$ is of order h^{n+1}, and so the most significant part of T_{i+1} must be attributed to $\overline{w}_{i+1} - w_{i+1}$. This means that we can write

$$T_{i+1} \approx \frac{1}{h}(\overline{w}_{i+1} - w_{i+1}) \tag{B.8}$$

and, since T_{i+1} is of order h^n, we have

$$kh^n \approx \frac{1}{h}\left(\overline{w}_{i+1} - w_{i+1}\right).$$ (B.9)

Thus we have in equaton (B.8) an estimation of the local truncation error for order h^n based on the numerical approximations generated by the method of order n and the method of order $n+1$. Equation (B.9) simply replaces T_{i+1} by kh^n where k is some constant. This will now be used to help estimate the step size.

Estimation of step size. Let us consider the local truncation error in equation (B.8) with h replaced by qh where q is bounded above and positive. This truncation error is denoted by $T_{i+1}(qh)$ and so equation (B.9) gives

$$T_{i+1}(qh) \approx k(qh)^n = q^n kh^n \approx \frac{q^n}{h}\left(\overline{w}_{i+1} - w_{i+1}\right).$$

Since we require $T_{i+1}(qh) \le \varepsilon$ for some specified $\varepsilon > 0$, we have

$$\frac{q^n}{h}\left|\overline{w}_{i+1} - w_{i+1}\right| \le \varepsilon$$

and so

$$q \le \left\{ \frac{\varepsilon h}{\left|\overline{w}_{i+1} - w_{i+1}\right|} \right\}^{1/n}$$ (B.10)

where n is the order of the local truncation error T_{i+1}.

The value of q is determined at each stage in the method, with the value at the i th stage having the following two purposes:

(i) to reject the initial choice of h_i at the i th step if necessary and repeat the calculations using qh_i,
(ii) to predict an appropriate initial choice of h_i for the $(i+1)$ th step.

Since there is a penalty in terms of function evaluations that must be paid if many of the steps are repeated, then in practice q tends to be chosen rather conservatively. In fact, for the Runge-Kutta-Fehlberg method discussed in detail in Section B.5 which uses the above error control theory, q is chosen to be

$$q = \left\{ \frac{\varepsilon h}{2|\overline{w}_{i+1} - w_{i+1}|} \right\}^{1/n} .$$

B.5 Runge-Kutta-Fehlberg Method

General introduction. One popular technique that uses the inequality (B.10) for solving the initial value problem is the Runge-Kutta-Fehlberg (R-K-F) method presented by Fehlberg [1] in 1970. This technique consists of using a Runge-Kutta method of order 5 to estimate the local truncation error in a Runge-Kutta method of order 4. A clear advantage of this method is that it only requires 6 functional evaluations per step, whereas with arbitary Runge-Kutta methods or orders 4 and 5 used together, we would require 10 functional evaluations per step.

The error control theory developed in Section B.4 leads to a procedure that requires twice the number of functional evaluations per step than a procedure without error control. Of course, it is important to include error control, but because of this penalty the value of q (see inequality (B.10)) tends to be chosen rather conservatively. In fact, for the R-K-F method of order 4(5), the usual choice of q is as follows:

$$q = \left\{ \frac{\varepsilon h}{2|\overline{w}_{i+1} - w_{i+1}|} \right\}^{1/4} = 0.84 \left\{ \frac{\varepsilon h}{|\overline{w}_{i+1} - w_{i+1}|} \right\}^{1/4} .$$

Computer algorithm. The algorithm below is based on the following equations presented by Fehlberg [1].

The problem is to solve the initial-value problem in equation (B.1) by means of Runge-Kutta methods of order 4 (w) and order 5 ($\overline{w}$). The step size which is variable is denoted by h, the tolerance which is the value the local truncation error is not to exceed is denoted by TOL and the number of iterations is denoted by the variable i.

$$w_0 = \alpha \quad \overline{w}_0 = \alpha \quad i = 0 \quad h = [\text{TOL}]^{1/4} \quad t = a$$

do

$$t = t + h$$
$$i = i + 1$$
$$w_i = w_{i-1} + \frac{25}{216}k_1 + \frac{1408}{2565}k_3 + \frac{2197}{4104}k_4 - \frac{1}{5}k_5$$
$$\overline{w}_i = \overline{w}_{i-1} + \frac{16}{135}k_1 + \frac{6656}{12825}k_3 + \frac{28561}{56430}k_4 - \frac{9}{50}k_5 + \frac{2}{55}k_6$$

if $|\overline{w}_i - w_i| = 0$ **then**

    $h = hmax$

else

$$q = 0.84 \left\{ \frac{\text{TOL}}{|\overline{w}_i - w_i|} \right\}^{1/4}$$

    if $qh < hmin$ **then**

        STOP – minimum step size exceeded

    else

        if $qh < h$ **then**

$$t = t - h + qh$$
$$h = qh$$
$$w_i = w_{i-1} + \frac{25}{216}k_1 + \frac{1408}{2565}k_3 + \frac{2197}{4104}k_4 - \frac{1}{5}k_5$$
$$\overline{w}_i = \overline{w}_{i-1} + \frac{16}{135}k_1 + \frac{6656}{12825}k_3 + \frac{28561}{56430}k_4 - \frac{9}{50}k_5 + \frac{2}{55}k_6$$

        else

            $h = qh$

        endif

    endif

  endif

until $t > b$

where

$$k_1 = hf\left(t, w_{i-1}\right)$$

$$k_2 = hf\left(t + \frac{1}{4}h, w_{i-1} + \frac{1}{4}k_1\right)$$

$$k_3 = hf\left(t + \frac{3}{8}h, w_{i-1} + \frac{3}{32}k_1 + \frac{9}{32}k_2\right)$$

$$k_4 = hf\left(t + \frac{12}{13}h, w_{i-1} + \frac{1932}{2197}k_1 - \frac{7200}{2197}k_2 + \frac{7296}{2197}k_3\right)$$

$$k_5 = hf\left(t + h, w_{i-1} + \frac{439}{216}k_1 - 8k_2 + \frac{3680}{513}k_3 - \frac{845}{4104}k_4\right)$$

$$k_6 = hf\left(t + \frac{1}{2}h, w_{i-1} - \frac{8}{27}k_1 + 2k_2 - \frac{3544}{2565}k_3 + \frac{1859}{4104}k_4 - \frac{11}{40}k_5\right)$$

The algorithm given above was combined into a FORTRAN computer program. The method is tested out in Section B.6 by applying it to the problem arising in the theory of contagious disease.

B.6 Solution of Contagious Disease Problem Using Runge-Kutta-Fehlberg Algorithm

The problem we need to solve is the differential equation given as equation (B.3). We need to give some initial values, however, since this is the nature of the problem. We will assume the number of infectives at time zero as numbering one thousand and the total population as a hundred thousand. We will also assume that the infection rate of the disease is 2×10^{-6}. What we require is the number of infectives at any particular time. We will run the program for one hundred days with a tolerance of one.

We therefore have the following equation to solve;

$$y' = y(n + a - y) \qquad y(0) = 1000.0 \qquad 0 \le t \le 100$$

where

n = initial number of susceptibles = 99000.0

a = initial number of infectives = 1000.0

β = infection (contact) rate = 2×10^{-6}

The program (as described previously in Section B.5) was run using a minimum step of 1/2 and a maximum step size of 1. Analytical and numerical results have been obtained at daily intervals over the time period $t = 0$ to 100 days. The following conclusion can be drawn from these results:

$$\text{Maximum absolute error} = 0.07948446 \quad (\text{at } t = 23)$$
$$\text{Maximum relative error} = 0.0002413\% \quad (\text{at } t = 5)$$

A summary showing the comparison of the computed and analytical results at 10 days intervals is given in Table B.1.

It is almost possible not to consider the rest of the data since the maximum relative error is so small, except to note that the majority of errors are smaller than even this figure. We also note that the errors are larger when the graph (shown in Figure B.1) is of steepest descent. Also, as the epidemic comes to its close; that is, as almost all the population becomes infected, the relative error approaches zero.

Although the program was run only for one hundred days, it can clearly be seen from the graph in Figure B.1 that the number of infectives in the population is approaching the total number of the population. This is what we concluded should happen (see Section B.2).

What we can deduce, therefore, from the reasons given above, as well as from the information given in Section B.5, is that the Runge-Kutta-Fehlberg method is a very good method for approximating the solution to an initial value problem. In this particular case, it approximates the solution very well indeed. We can also be reasonably safe in assuming that it would approximate equally well the same problem, but with different initial values as the method is stable.

A plot of the computed solution is given in Figure B.1. This shows the number of infectives in the population over a 100 day period. The actual solution is plotted on the same graph but coincides with the numerical solution as the two can not be distinguished.

t	Numerical solution w_i $(\times 10^4)$	Analytical solution y_i $(\times 10^4)$	Absolute error $\lvert y_i - w_i \rvert$ $(\times 10^{-4})$	Percentage relative error $(\times 10^{-5})$
0	0.1000 0000	0.1000 0000	0.00	0.00
10	0.6945 3303	0.6945 3160	142.97	20.59

20	3.5546 1713	3.5546 0987	725.90	20.42
30	8.0295 7681	8.0295 7153	528.71	6.58
40	9.6785 6906	9.6785 6704	201.56	2.08
50	9.9552 5571	9.9552 5518	53.36	0.54
60	9.9939 2104	9.9939 2093	10.89	0.11
70	9.9991 7687	9.9991 7685	1.95	0.02
80	9.9998 8859	9.9998 8859	0.31	0.00
90	9.9999 8492	9.9999 8492	0.04	0.00
100	9.9999 9796	9.9999 9796	0.01	0.00

Table B.1. Comparison of numerical and analytical results.

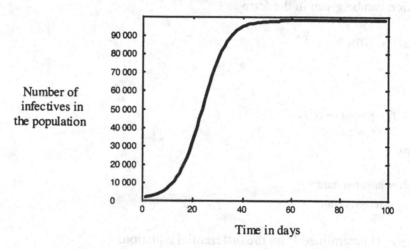

Figure B.1. Plot of the number of infectives in the population against time in days.

Clearly we see from Table B.1 that there is very close agreement between the actual solution y_i and the computed solution w_i. This is apparent when examining the values of the error $w_i - y_i$ and the percentage relative error. From these it is possible to find an approximation to the number of infectives at any one time. For example, the number of infectives in the population after 30 days would be 80296.

B.7 Conclusion

The case study has considered a deterministic model in the theory of contagious disease. For this particular case we have assumed that all infected individuals remain in the population to spread the disease.

From the numerical study we can conclude that the Runge-Kutta-Fehlberg method is a particularly good method for solving initial value problems of this type in that it uses error control, is stable and converges to the real solution as the step size tends to zero.

Fortunately for this particular mathematical model we have been able to compare computed and exact solutions. A more realistic situation is obtained by introducing a third variable $z(t)$ to represent the number of individuals who are removed from the affected population at a given time t, either by isolation, recovery and consequent immunity, or death. This quite naturally complicates the problem, but it can be shown that an approximate solution can be given in the form

$$x(t) = x(0)e^{-(k_1/k_2)z(t)}$$

and

$$y(t) = n - x(t) - z(t),$$

where

$$k_1 = \text{infective rate}$$
$$k_2 = \text{removal rate},$$

and $z(t)$ is determined from the differential equation,

$$z'(t) = k_2\left\{n - z(t) - x(0)e^{-(k_1/k_2)z(t)}\right\}.$$

This particular problem can not be solved directly and so a numerical procedure based on the method presented in this case study would need to be applied.

B.8 Problems

1. Extend the contagious disease case study to take into account removal of infectives from circulation by death or isolation. Make the following modelling assumptions:
(1) the removals include infectives who are isolated, dead or recovered and immune;
(2) the immune or recovered removals enter a new category which is not

susceptible to the disease.

Suppose we have a group of total size n of whom, at time t, there are x susceptible, y infectives and z individuals who are isolated, dead or recovered and immune. Thus $x + y + z = n$. Assuming

β = contact rate,

γ = rate of removal of infectives from circulation,

model the deterministic process by a system of 3 first order differential equations in x, y and z with initial conditions $(x, y, z) = (x_0, y_0, 0)$ when $t = 0$. Define a relative removal rate given by $\rho = \gamma/\beta$. Thus the value $\rho = x_0 \approx n$ constitutes a threshold density of susceptibles.

2. Obtain an approximate solution to the mathematical model in Problem 1 by undertaking the following steps:

(a)   Show that

$$\frac{dz}{dt} = \gamma\left(n - z - x_0 e^{-z/\rho}\right).$$

(b)   Further, show that up to terms in z^2 this can be approximated by

$$\frac{dz}{dt} = \gamma\left\{n - x_0 + \left(\frac{x_0}{\rho} - 1\right)z - \frac{x_0}{2\rho^2}z^2\right\}.$$

(c)   Hence, show that the epidemic curve is given by

$$\frac{dz}{dt} = \frac{\gamma\alpha^2\rho^2}{2x_0}\operatorname{sech}^2\left(\frac{1}{2}\alpha\gamma t - \psi\right),$$

where

$$\alpha = \left\{\left(\frac{x_0}{\rho} - 1\right)^2 + \frac{2x_0 y_0}{\rho^2}\right\}^{1/2},$$

and

$$\psi = \tanh^{-1}\left(\frac{x_0 - \rho}{\alpha\rho}\right).$$

(d)   Comment on the shape of the graph and relate this to an actual epidemic situation.

(e)   If $x_0 \approx n$, find the final value z_∞ of z when the epidemic is over. Comment on total size of the epidemic, i.e., the total number of removals after a very long period of time.

3. Develop a mathematical model for a population which is subject to a disease which is seldom fatal and leaves the victim immune to future infections by the disease. Assume that infection can only occur when a susceptible person comes in direct contact with an infectious person.

In the model assume the population has 100000 members and the infectious period lasts approximately 3 weeks. Also last week there were 18 new cases of the disease reported. This week there were 40 new cases. It is estimated that 30% of the population is immune due to previous exposure.
(a)   Using the above data, find the eventual number of people who will become infected.
(b)   Estimate the maximum number of new cases in any one week.
(c)   Conduct a sensitivity analysis to investigate the effect of any assumptions you make in part (a) which are not supported by real data.
(d)   Perform a sensitivity analysis for the number (18) of cases reported last week. It is thought by some that in early weeks the epidemic might be underreported.

Bibliography

1.   Fehlberg, E., "Klassische Runge-Kutta-Formeln vierter und niedrigerer Ordnung mit Schrittweiten-Kontrolle und ihre Anwendung auf Wärmeleitungsprobleme", Computing, 1970, **6**, 61–71.
2.   Bailey, N.T.J., *The Mathematical Approach to Biology and Medicine*, John Wiley & Sons, London, 1967.
3.   Bailey, N.T.J., *The Mathematical Theory of Epidemics*, C. Griffin, London, 1957.
4.   Maron, M.J., *Numerical Analysis. A Practical Approach*, Macmillan, New York, 1982.
5.   Gerald, C.F. and Wheatley, P.O., *Applied Numerical Analysis*, 3rd ed., Addison-Wesley, Reading, MA, 1984.
6.   Rice, J.R., *Numerical Methods. Software and Analysis*, McGraw-Hill, New York, 1983.

Case Study C

NUMERICAL CAM DESIGN

C.1 The Cam-Follower Mechanism

The function of the disk cam with the flat-faced radial follower shown in Figure C.1 is to transform the angular motion of the disk cam to a linear motion of the follower. A basic design problem is to determine the contour of the disk cam that generates a prescribed follower motion. Minimisation of the cam dimensions is another important design consideration. During the motion of the cam-follower mechanism the cam slides against the follower face and in order to avoid rapid wear the disk cam is required to be smooth. The avoidance of cusps in the cam is thus a design constraint that must be achieved.

In practice the disk cam rotates and the follower motion is linear. To simplify the analysis, however, we follow Carver and Quinn [1] and invert the mechanism, hold the disk cam stationary and rotate the follower about it, as shown in Figure C.2. We denote the minimal radius of the cam by c and assume that initially the mechanism is in the position shown in Figure C.1(a). Then, for an angle of rotation θ we denote

$$r = c + f(\theta) \tag{C.1}$$

where $f(\theta)$ is the follower motion. Inspecting Figure C.2 we see that the face of the follower satisfies the straight line equation

$$y = -x \cot \theta + r \csc \theta . \tag{C.2}$$

Hence, substituting (C.2) in (C.1) we have

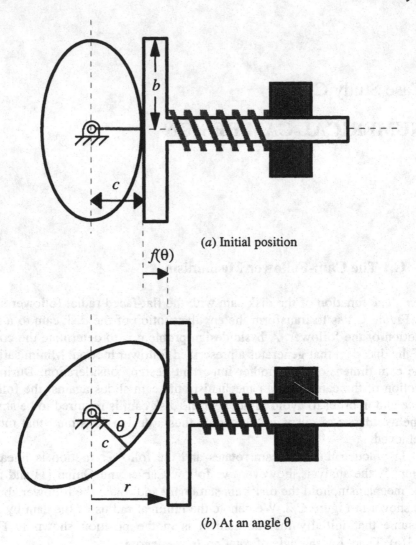

(a) Initial position

$f(\theta)$

(b) At an angle θ

Figure C.1 Motion of the cam-follower mechanism

$$c + f(\theta) = x\cos\theta + y\sin\theta. \tag{C.3}$$

Differentiating (C.3) gives

$$f'(\theta) = -x\sin\theta + y\cos\theta \tag{C.4}$$

where prime denotes derivative with respect to θ. Solving (C.3) and (C.4) for x and y yields the parametric equations for the contour of the disk cam

$$x = (c + f)\cos\theta - f'\sin\theta,$$ (C.5)

$$y = (c + f)\sin\theta + f'\cos\theta.$$ (C.6)

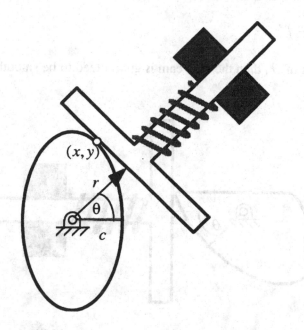

Figure C.2 Follower motion with respect to the disk

If the disk cam is not smooth then the cusp may remain the point of contact with the follower face while θ changes through some finite interval, as in the position shown in Figure C.3. It thus follows that cusp occurs if both

$$\frac{dx}{d\theta} = \frac{dy}{d\theta} = 0$$ (C.7)

for some angle θ. Differentiating (C.5) and (C.6) with respect to θ yields

$$\frac{dx}{d\theta} = -(c + f + f'')\sin\theta$$ (C.8)

and

$$\frac{dy}{d\theta} = (c + f + f'')\cos\theta.$$ (C.9)

Therefore, cusp occurs when $c + f + f'' = 0$ for some values of θ. If we ensure however that c is large enough such that

$$c > -f - f''$$ (C.10)

for all values of θ, then the disk cam is guaranteed to be smooth.

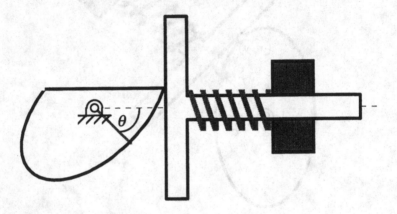

Figure C.3 Diagram showing position of cusp and follower face

The distance L between the centre axis of the follower and the point of contact is

$$L = \sqrt{(r\cos\theta - x)^2 + (y - r\sin\theta)^2}$$ (C.11)

Substituting (C.5) and (C.6) in (C.11) gives

$$L = f'.$$ (C.12)

It follows therefore that in order to ensure contact between the face of the follower and the disk cam we must ensure that the length b of the follower face satisfies

$$b > \max|f'|.$$ (C.13)

C.2 Test Example

Design a disk cam that produces the motion (in millimetres)

$$f = 0.0001\theta^8(\theta - 2\pi)^2, \ 0 \le \theta \le 2\pi, \tag{C.14}$$

of the follower.

Solution
Differentiating f we have

$$f' = 0.0001(10\theta^9 - 36\pi\theta^8 + 32\pi^2\theta^7), \tag{C.15}$$

$$f'' = 0.0001(90\theta^8 - 288\pi\theta^7 + 224\pi^2\theta^6). \tag{C.16}$$

The follower displacement and its first two derivatives are shown in Figure C.4.

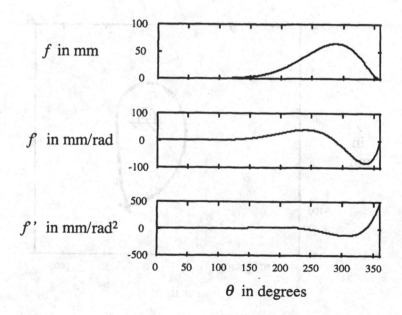

Figure C.4 Follower displacement and its first two derivatives

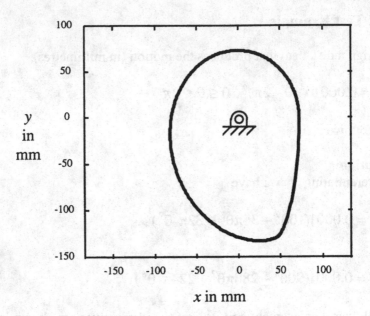

Figure C.5 A disk cam satisfying the no-cusp condition (C.10)

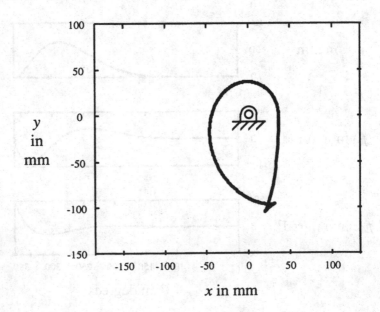

Figure C.6 Contour violating the no-cusp condition (C.10)

To satisfy the no-cusp condition (C.10) we choose

$$c = -1.05 \min(f + f') = 72.81 \, \text{mm} \tag{C.17}$$

and obtain, using the parametric equations (C.5), (C.6), the disk cam shown in Figure C.5.

We now demonstrate the importance of the no-cusp condition. In violation of (C.10) a minimal cam radius of

$$c = -0.5 \min(f + f') = 36.41 \, \text{mm} \tag{C.18}$$

is drawn in Figure C.6. Clearly, apart from cusps the parametric equations (C.5) and (C.6) do not produce a physically realisable disk cam in this case.

C.3 Discrete Follower Motion

In some applications the follower motion data is given in terms of numerical values and not by an analytical function $f(\theta)$. Moreover, measurement data are usually noise corrupted. Suppose that the follower motion (C.14) has been sampled with uniform sampling rate $\Delta\theta = \pi / 180$ rad. With measurement error e_i for the i-th data point we have

$$f_i = 0.0001(i\Delta\theta)^8 (i\Delta\theta - 2\pi)^2 + e_i , \; i = 1,2,...,360. \tag{C.19}$$

We have used random data uniformly distributed in the interval [0,1] to represent the noise e_i.

The derivatives of the discrete function f_i may be found by numerical differentiation

$$f_i' = \frac{f_{i+1} - f_{i-1}}{2\Delta\theta}, \; i = 1,2,...,360 \tag{C.20}$$

$$f_i'' = \frac{f_{i+1} - 2f_i + f_{i-1}}{(\Delta\theta)^2}, \; i = 1,2,...,360 \tag{C.21}$$

where, due to the periodicity of f_i,

$$f_0 = f_{360} \text{ and } f_{361} = f_1. \tag{C.22}$$

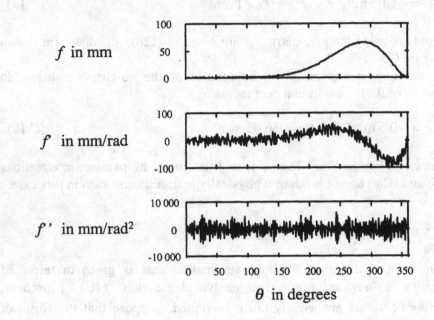

Figure C.7 Follower displacement and its first two derivatives for noisy data

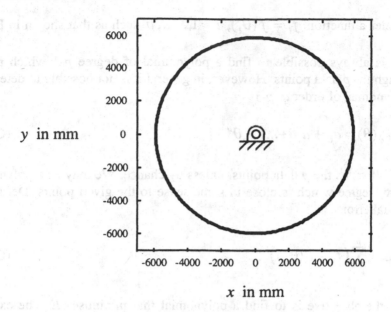

Figure C.8 Follower contour for noisy data

The follower displacement for this case, shown in Figure C.7, appears to be similar to that in Figure C.4. Its first derivative however has been corrupted significantly by noise. The second derivative looks like a random series. Correspondingly, the contour of the cam produced from these data, shown in Figure C.8, looks different from its counterpart in Figure C.5. It is much larger that the cam obtained using the smooth follower displacement.

We have just observed the phenomenon known as *catastrophic cancellation*. When differentiating a function consisting of signal and noise, the majority of the signal cancels out while the noise may build up. The noise is just magnified by the process of numerical differentiation. For higher derivatives the noise is magnified further. It thus follows that numerical differentiation should be avoided when dealing with noise corrupted data. We may smooth the data and partially overcome the difficulty observed, as demonstrated in the following section.

C.4 Least-Squares Smoothing

The idea of the least square smoothing method is to approximate the noisy data by a polynomial of low degree. In order to establish this concept consider a function $f_i = f(\theta_i)$, $i = 1,2,...,p$ such as that shown in Figure C.9.

It is always possible to find a polynomial of degree $p\text{-}1$ which passes through the p data points. However, in general it is not possible to determine a polynomial of order $q<p\text{-}1$

$$P_q(\theta) = a_0 + a_1\theta + ... + a_q\theta^q \qquad (C.23)$$

that collocates the p data points, unless by chance. We may fit a polynomial of low degree which is close in some sense to the given points. Define the residual error

$$E = \sum_{k=1}^{p}(f_k - P(\theta_k))^2. \qquad (C.24)$$

Then the objective is to find a polynomial that minimises E. The extreme value of E may be determined by solving

$$\frac{\partial E}{\partial a_j} = 0, \quad j = 0,1,...,q. \qquad (C.25)$$

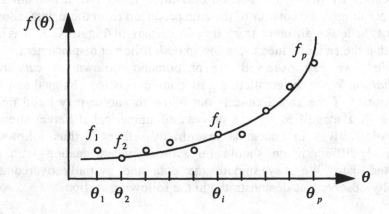

Figure C.9 Least squares interpolation

It is clear from physical considerations that E is unbounded. Since E is also non-negative it follows that the extreme value of E is a minimum point.

Substituting (C.24) in (C.25) and differentiating, we obtain for $j = 0,1,...,q$,

$$2\sum_{k=1}^{p}\left(f_k - (a_0 + a_1\theta_k + ... + a_q\theta_k^q)\right)\theta_k^j = 0 \qquad (C.26)$$

or

$$a_0\sum_{k=1}^{p}\theta_k^j + a_1\sum_{k=1}^{p}\theta_k^{j+1} + ... + a_q\sum_{k=1}^{p}\theta_k^{j+q} = \sum_{k=1}^{p}\theta_k^j f_k . \qquad (C.27)$$

The $q+1$ equations (C.27) for $j = 0,1,2,...,q$ can be written in matrix form

$$\begin{bmatrix} p & \sum_{k=1}^{p}\theta_k & \cdots & \sum_{k=1}^{p}\theta_k^q \\ \sum_{k=1}^{p}\theta_k & \sum_{k=1}^{p}\theta_k^2 & \cdots & \sum_{k=1}^{p}\theta_k^{q+1} \\ \vdots & \vdots & \cdots & \vdots \\ \sum_{k=1}^{p}\theta_k^q & \sum_{k=1}^{p}\theta_k^{q+1} & \cdots & \sum_{k=1}^{p}\theta_k^{2q} \end{bmatrix} \begin{pmatrix} a_0 \\ a_1 \\ \vdots \\ a_q \end{pmatrix} = \begin{pmatrix} \sum_{k=1}^{p}f_k \\ \sum_{k=1}^{p}\theta_k f_k \\ \vdots \\ \sum_{k=1}^{p}\theta_k^q f_k \end{pmatrix}. \qquad (C.28)$$

Note that if we define the $p \times (q+1)$ *Vandermonde* matrix

$$\mathbf{V} = \begin{bmatrix} 1 & \theta_1 & \cdots & \theta_1^q \\ 1 & \theta_2 & \cdots & \theta_2^q \\ \vdots & \vdots & \cdots & \vdots \\ 1 & \theta_p & \cdots & \theta_p^q \end{bmatrix}, \qquad (C.29)$$

then the normal equations (C.28) may be written in the convenient form

$$\mathbf{V}^T\mathbf{V}\mathbf{a} = \mathbf{V}^T\mathbf{f} \qquad (C.30)$$

where

$$\mathbf{a} = (a_0, a_1, ..., a_q)^T \text{ and } \mathbf{f} = (f_1, f_2, ..., f_p)^T .$$

We now consider the case where a second degree polynomial is fitted to five equally spaced data points as shown in Figure C.10. The normal equations for this case are

$$
\begin{bmatrix}
5 & 0 & 10(\Delta\theta)^2 \\
0 & 10(\Delta\theta)^2 & 0 \\
10(\Delta\theta)^2 & 0 & 34(\Delta\theta)^4
\end{bmatrix}
\begin{pmatrix} a_0 \\ a_1 \\ a_2 \end{pmatrix}
$$

$$
= \begin{pmatrix}
f_{-2} + f_{-1} + f_0 + f_1 + f_2 \\
(-2f_{-2} - f_{-1} + f_1 + 2f_2)\Delta\theta \\
(4f_{-2} - f_{-1} + f_1 + 4f_2)(\Delta\theta)^2
\end{pmatrix}
\tag{C.31}
$$

which yields the solution

$$
a_0 = \frac{-3f_{-2} + 12f_{-1} + 17f_0 + 12f_1 - 3f_2}{35},
\tag{C.32}
$$

$$
a_1 = \frac{-2f_{-2} - f_{-1} + f_1 + 2f_2}{10\Delta\theta},
\tag{C.33}
$$

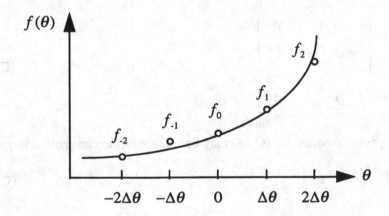

Figure C.10 Second degree polynomial fitted to five points

$$a_2 = \frac{2f_{-2} - f_{-1} - 2f_0 - f_1 + 2f_2}{14(\Delta\theta)^2}. \tag{C.34}$$

The quadratic polynomial that fits the five equally spaced points is thus

$$P_2(\theta) = a_0 + a_1\theta + a_2\theta^2 \tag{C.35}$$

where the coefficients a_0, a_1 and a_2 are given by (C.32) to (C.34).

In particular, substituting for $\theta = 0$ in (C.35), the values of f_0 may be replaced by the smoothed value

$$\hat{f}_0 = a_0. \tag{C.36}$$

More generally, in view of (C.36) and (C.32), we conclude that the i-th value $\hat{f}_i$ of a smoothed function may be approximated by the values of its neighbouring points $f_{i-2}, f_{i-1}, f_i, f_{i+1}$ and f_{i+2} according to the smoothing formula

$$\hat{f}_i = \frac{-3f_{i-2} + 12f_{i-1} + 17f_i + 12f_{i+1} - 3f_{i+2}}{35}. \tag{C.37}$$

We may approximate the value of the first derivative of f by using

$$\hat{f}_0' = P_2'(0) = a_1. \tag{C.38}$$

It thus follows by (C.33) that the smoothed value of f_i' is given by

$$\hat{f}_i' = \frac{-2f_{i-2} - f_{i-1} + f_{i+1} + 2f_{i+2}}{10\Delta\theta}. \tag{C.39}$$

Similarly, from

$$\hat{f}_0'' = P_2''(0) = 2a_2 \tag{C.40}$$

we find that f_i'' may be approximated by

$$\hat{f}_i'' = \frac{2f_{i-2} - f_{i-1} - 2f_i - f_{i+1} + 2f_{i+2}}{7(\Delta\theta)^2}. \tag{C.41}$$

C.5 Discrete Follower Motion Revisited

We consider again the example of Section C.3 and use the smoothing formula (C.37) to smooth the follower displacement. We have applied the smoothing formula successively 100 times to obtain a reasonably smoothed function $f(\theta)$. Then the numerical differentiations have been carried out. The results are displayed in Figure C.11. The disk cam which has been derived by using the smoothed data is shown in Figure C.12. Now it resembles the originally designed cam (see Figure C.5).

It should be noted, however, that such a remarkable recovery of a signal from noisy data cannot be expected in real life problems. Real measurement errors usually depend heavily on *systematic errors*.

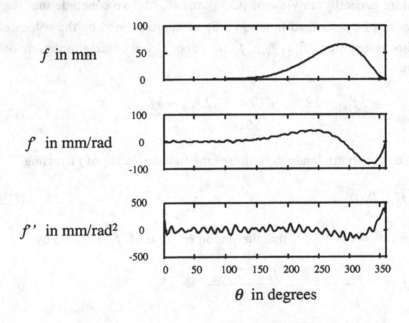

Figure C.11 Results obtained by numerical differentiation on smoothed data

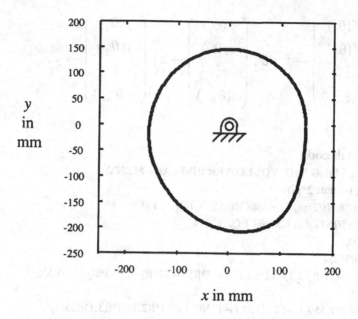

Figure C.12. The disk cam derived by using the smoothed data

C.6 Matlab Program

The following Matlab program produces the disk cam for the case studied in Section C.5. Explanatory comments, not to be executed by the program, start with % sign. The following variables and arrays have been used in the computer program.

Constants and Variables.

$\Delta\theta \rightarrow dt$; $c \rightarrow c$; j,k integers.

Arrays

$$\begin{pmatrix} \theta_1 \\ \theta_2 \\ \vdots \\ \theta_{360} \end{pmatrix} \rightarrow t \; ; \quad \begin{pmatrix} f(\theta_1) \\ f(\theta_2) \\ \vdots \\ f(\theta_{360}) \end{pmatrix} \rightarrow f \; ; \quad \begin{pmatrix} f'(\theta_1) \\ f'(\theta_2) \\ \vdots \\ f'(\theta_{360}) \end{pmatrix} \rightarrow \text{f\_d} \; ;$$

$$
\begin{pmatrix} f''(\theta_1) \\ f''(\theta_2) \\ \vdots \\ f''(\theta_{360}) \end{pmatrix} \rightarrow \text{f\_dd} \; ; \quad
\begin{pmatrix} x(\theta_1) \\ x(\theta_2) \\ \vdots \\ x(\theta_{360}) \end{pmatrix} \rightarrow \text{x} \; ; \quad
\begin{pmatrix} y(\theta_1) \\ y(\theta_2) \\ \vdots \\ y(\theta_{360}) \end{pmatrix} \rightarrow \text{y} .
$$

The Matlab code

```
% GENERATING THE FOLLOWER DISPLACEMENT
dt=pi/180; t=(dt:dt:2*pi)';
f=0.0001*t.^8.*(t-2*pi).^2+rand(360,1); % Eq. (C.19)
% SMOOTHING f(θ) USING EQ. (C.37)
for k=1:100,
    for j=3:358,
        f(j,1)=(-3*f(j-2,1)+12*f(j-1,1)+17*f(j,1)+12*f(j+1,1)-3*f(j+2,1))/35;
        end;
    f(1,1)=(-3*f(359,1)+12*f(360,1)+17*f(1,1)+12*f(2,1)-3*f(3,1))/35;
    f(2,1)=(-3*f(360,1)+12*f(1,1)+17*f(2,1)+12*f(3,1)-3*f(4,1))/35;
    f(360,1)=(-3*f(358,1)+12*f(359,1)+17*f(360,1)+12*f(1,1)-3*f(2,1))/35;
    f(359)=(-3*f(357,1)+12*f(358,1)+17*f(359,1)+12*f(360,1)-3*f(1,1))/35;
    end;
% DETERMINING f'(θ)
for j=2:359,
    f_d(j,1)=(f(j+1,1)-f(j-1,1))/(2*dt); % Eq. (C.20)
    end;
f_d(1,1)=(f(2,1)-f(360,1))/(2*dt);
f_d(360,1)=(f(1,1)-f(359,1))/(2*dt);
% DETERMINING f''(θ)
for j=2:359,
    f_dd(j,1)=(f(j+1,1)-2*f(j,1)+f(j-1,1))/dt^2; % Eq. (C.21)
    end;
f_dd(1,1)=(f(2,1)-2*f(1,1)+f(360,1))/dt^2;
f_dd(360,1)=(f(1,1)-f(360,1)+f(359,1))/dt^2;
% DETERMINING THE CONTOUR OF THE CAM
c=-1.05*min(f+f_dd); % Eq. (C.17)
x=(c+f).*cos(t)-f_d.*sin(t); % Eq. (C.5)
y=(c+f).*sin(t)+f_d.*cos(t); % Eq. (C.6)
% PLOTTING THE RESULTS
plot(x,y)
```

C.7 Problems

1. Design and plot a disk cam that while rotating at a constant angular velocity produces the follower motion (in millimetres)

$$f(\theta) = 0.01\theta^2 (2\pi - \theta)^4 \sin^2\theta, \ 0 \le \theta \le 2\pi,$$

where θ is the angle of rotation of the cam.

2. It is required that the slider B shown in Figure C.13 will move out and back according to the equation

$$z = 60(1 - \cos(50t)),$$

where z is the displacement in millimetres of the slider B, and t is the time in seconds. Design and plot the disk cam that rotates with constant angular velocity and produces this motion.

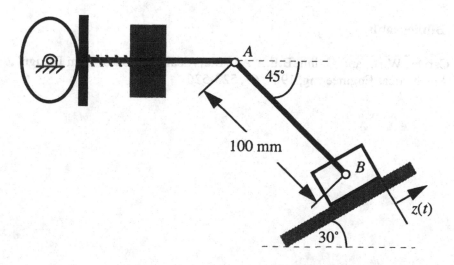

FigureC.13 A cam-follower-slider mechanism

3. Design and plot a disk cam which produces the motion

$$\psi = \frac{\pi}{12}(1 - \cos\theta), \ 0 \le \theta \le 2\pi,$$

for the rod AB shown in Figure C.14, where θ is the angle of rotation of the cam and ψ is the angle of AB, both measured in radians.

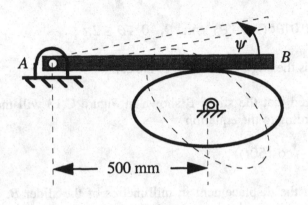

Figure C.14 A cam-rod mechanism

Bibliography

1. Carver, W.B., and Quinn, B.E., "An Analytical Method of Cam Design", Mechanical Engineering, 1945, **67**, 523-526

Case Study D

DYNAMICS AND FORCE ANALYSIS OF A MECHANISM

D.1 The Slider-Crank Mechanism

The slider-crank mechanism, shown in Figure D.1, consists of two bars and a sliding block. Its components, the bars and the block, are called links. For the sake of definiteness we number the links as shown in the figure. Note that the ground, which may also be regarded as a link, is called link 1. It is in contact with link 2 via the pin A and with link 4 through the sliding guide. Pin B connects link 2 to link 3, and pin C connects link 3 to link 4. The purpose of this mechanism is to transmit the angular motion of link 2 to the linear motion of sliding link C, and *vice versa*. Such a mechanism forms the basic component in the internal combustion engine, for example.

An external moment M_2 is applied to link 2, as shown in Figure D.1, which derives the motion of the mechanism. Denote the angle between link 2 and the horizontal by θ_2. Note that the sense of increasing θ_2 is chosen such that an anti-clockwise motion of link 2 about A is characterised by positive angular velocity $\dot{\theta}_2$. For clockwise motion of link 2 about A we have $\dot{\theta}_2 < 0$.

In this case study we will address the following problems:

Problem 1. Determine the external moment M_2 required to derive the mechanism such that link 2 rotates with a given constant speed.

Problem 2. Determine the forces developed in pins A, B and C.

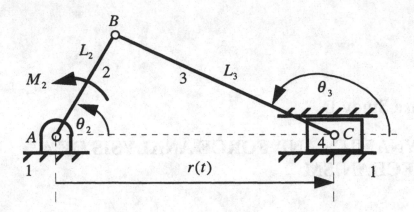

Figure D.1 The slider-crank mechanism

Note that only link 2 moves with constant angular velocity. The angular velocity of link 3 and the linear velocity of slider 4 are generally not constant. It is thus expected that the external moment M_2 which solves Problem 1 is time variable and can be expressed as $M_2 = M_2(\theta_2)$. Similarly, the pin forces are also functions of θ_2.

For numerical values we use
    Length: $L_2 = 4$ m, $L_3 = 8$ m.
    Mass: $m_2 = 10$ Kg, $m_3 = 20$ Kg, $m_4 = 100$ Kg.
    Moment of inertia about G: $I_2 = 15$ Kg·m², $I_3 = 120$ Kg·m².
    Angular velocity: $\dot{\theta}_2 = 10$ rad/s.

We further assume that link 2 and link 3 are uniform, that pin C passes through the center of mass of slider 4, and that the sliding guide is smooth, ie. that no friction force is applied to the slider 4.

The mechanism is a multi-body dynamic system with constraints. The equations of motion for each link are given by equation (3.8) in Chapter 3. In our case the motion is prescribed and we are required to determine the forces and the applied moment. In Section D.2 we will determine the position of block C and the angular position of link 3 as functions of θ_2. Then, the velocity and acceleration of the mechanism will be determined in Sections D.3 and D.4. The required moment and forces will then be found by using (3.8) in Section D.5. The output will give us the forces and their variations applied to the pins. This information is essential in designing the mechanism appropriately to prevent mechanical failure.

For a mechanism in rapid motion, inertia forces are usually much larger than forces applied by gravity. For simplicity we will therefore ignore the effect of gravity on the forces developed in the pin connections.

D.2 The Position of the Mechanism

We define the angle θ_3 with positive anti-clockwise sense, as shown in Figure D.1. Then expressing the y-coordinate of B in terms of the two links and their angles gives

$$L_2 \sin \theta_2 = L_3 \sin \theta_3 . \tag{D.1}$$

Similarly the x-coordinate of B gives

$$L_2 \cos\theta_2 = r + L_3 \cos\theta_3 . \tag{D.2}$$

We may thus determine the angle θ_3 from (D.1)

$$\theta_3 = \sin^{-1}\left(\frac{L_2}{L_3} \sin\theta_2 \right) \tag{D.3}$$

and then find the slider position $r(\theta_2)$ from (D.2).

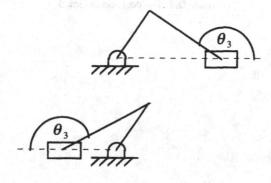

Figure D.2 Two possible configurations of the mechanism for θ_2

There are two possible solutions to (D.3), θ_3 and $\pi - \theta_3$. They correspond to the two possible configurations of the slider-crank mechanism of links L_2 and L_3 at the position θ_2, shown in Figure D.2. The condition

$$\frac{\pi}{2} < \theta_3 < \frac{3\pi}{2}, \tag{D.4}$$

which holds in our case, allows us to determine θ_3 uniquely from (D.3). The angular position of link 3 and the location of slider 4 are shown in Figures D.3 and D.4, respectively.

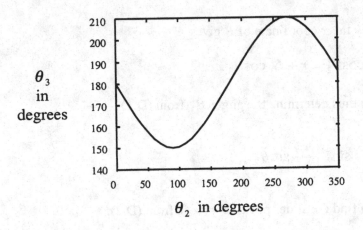

Figure D.3 The position of link 3

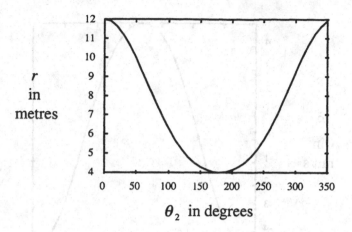

Figure D.4 The position of slider *C*

D.3 Velocity Analysis

We differentiate equations (D.1) and (D.2) with respect to time using the chain rule

$$\frac{d \sin\theta}{dt} = \frac{d \sin\theta}{d\theta} \frac{d\theta}{dt} = \dot{\theta}\cos\theta \, , \tag{D.5}$$

$$\frac{d \cos\theta}{dt} = \frac{d \cos\theta}{d\theta} \frac{d\theta}{dt} = -\dot{\theta}\sin\theta \, , \tag{D.6}$$

and obtain

$$L_2\dot{\theta}_2 \cos\theta_2 = L_3\dot{\theta}_3 \cos\theta_3 \tag{D.7}$$

$$-L_2\dot{\theta}_2 \sin\theta_2 = \dot{r} - L_3\dot{\theta}_3 \sin\theta_3 \, . \tag{D.8}$$

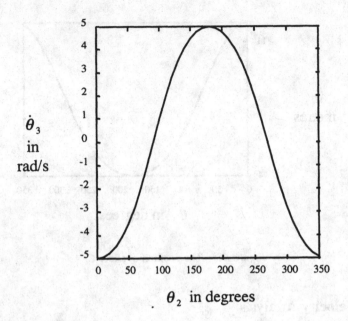

Figure D.5 Angular velocity of link 3

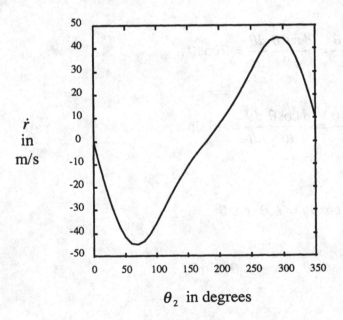

Figure D.6 Velocity of slider *C*

We recall that $\dot{\theta}_2$ is given. So $\dot{\theta}_3$ can be obtained from (D.7) and then $\dot{r}$ is determined by (D.8). This defines the velocity of the slider and the angular velocity of link 3. They are plotted in Figures D.5 and D.6.

D.4 Acceleration Analysis

In order to determine the acceleration of the slider and the angular acceleration of link 3 we differentiate (D.7) and (D.8) with respect to time, using the chain rule again, and obtain

$$L_2\ddot{\theta}_2\cos\theta_2 - L_2\dot{\theta}_2^2\sin\theta_2 = L_3\ddot{\theta}_3\cos\theta_3 - L_3\dot{\theta}_3^2\sin\theta_3, \qquad (D.9)$$

$$-L_2\ddot{\theta}_2\sin\theta_2 - L_2\dot{\theta}_2^2\cos\theta_2 = \ddot{r} - L_3\ddot{\theta}_3\sin\theta_3 - L_3\dot{\theta}_3^2\cos\theta_3. \qquad (D.10)$$

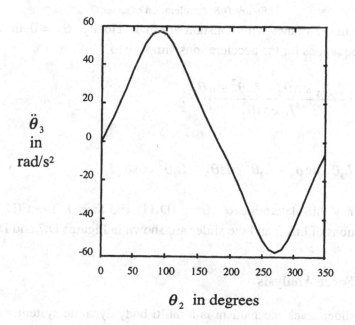

$\ddot{\theta}_3$
in
rad/s²

θ_2 in degrees

Figure D.7 Angular acceleration of link 3

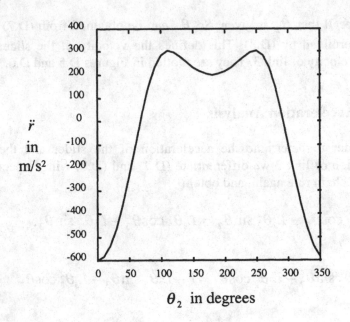

$\ddot{r}$
in
m/s²

θ_2 in degrees

Figure D.8 Acceleration of slider *C*

Since link 2 rotates with constant angular velocity, $\ddot{\theta}_2 = 0$ in our case, and the equations for the accelerations simplify to

$$\ddot{\theta}_3 = \frac{L_3 \dot{\theta}_3^2 \sin\theta_3 - L_2 \dot{\theta}_2^2 \sin\theta_2}{L_3 \cos\theta_3}, \tag{D.11}$$

$$\ddot{r} = L_3 \ddot{\theta}_3 \sin\theta_3 + L_3 \dot{\theta}_3^2 \cos\theta_3 - L_2 \dot{\theta}_2^2 \cos\theta_2. \tag{D.12}$$

We may thus determine $\ddot{\theta}_3$ from (D.11) and then $\ddot{r}$ from (D.12). The accelerations of link 3 and the slider are shown in Figures D.7 and D.8.

D.5 Force Analysis

The slider-crank mechanism is a multi body dynamic system, consisting of two links and a sliding block. Free body diagrams for its components are shown in Figure D.9. The force that link *i* applies on link *j* is denoted by F_{ij}, its *x*-component is X_{ij}, the *y*-component Y_{ij}. The positive *x* and *y* senses are right and up, respectively, and anti-clockwise is the sense of a positive moment.

Hence, the free body diagram of link 2 consists of the two components of pin force A, X_{12} and Y_{12}, which the ground (link 1) applies on link 2, other two components of pin force B, X_{32} and Y_{32}, which link 3 applies on link 2, and the external moment M_2 that derives the motion of the mechanism. They are drawn in the positive sense, as shown in the free body diagram for link 2. The centre of mass of link 2 is denoted by P. Hence, invoking (3.8) we have the following three equations associated with link 2

$$X_{12} + X_{32} = m_2 \ddot{x}_P \qquad (D.13)$$

$$Y_{12} + Y_{32} = m_2 \ddot{y}_P \qquad (D.14)$$

$$X_{12} \frac{L_2 \sin\theta_2}{2} - Y_{12} \frac{L_2 \cos\theta_2}{2} - X_{32} \frac{L_2 \sin\theta_2}{2}$$
$$+ Y_{32} \frac{L_2 \cos\theta_2}{2} + M_2 = I_2 \ddot{\theta}_2. \qquad (D.15)$$

Note that the moment due to the X_{12} component of pin force A is positive, as its effect is to rotate the link about the mass centre P in the positive anti-clockwise sense. The magnitude of this moment equals to the product of the force component X_{12} and its arm

$$\frac{L_2 \sin\theta_2}{2},$$

ie. the distance from P to X_{12}. By similar reasoning the moments associated with Y_{12}, X_{32} are negative, while the moment corresponding to Y_{32} is positive. The arms of Y_{12}, X_{32} and Y_{32} about the mass center P are

$$\frac{L_2 \cos\theta_2}{2}, \frac{L_2 \sin\theta_2}{2} \quad \text{and} \quad \frac{L_2 \cos\theta_2}{2},$$

respectively, and this is how (D.15) has been derived.

Two forces are applied on link 3 by the two pins B, and C. Using our notation their x- and y-components are X_{23}, Y_{23}, X_{43} and Y_{43}. By Newton's third law (3.6)

$$X_{23} = -X_{32}, Y_{23} = -Y_{32}$$

since the force that link 2 applies on link 3 is equal in magnitude and opposite in direction to the force that link 3 applies on link 2. To avoid the introduction of new unknowns X_{23}, Y_{23}, we have drawn in the free body diagram for link 3 the components $-X_{32}$ and $-Y_{32}$ which have been introduced already in (D.13) to (D.15). Accordingly, they appear in the negative sense, while the components of pin force C are assumed to be positive. The centre of mass for link 3 is denoted by Q, and from (3.8) we have

$$-X_{32} + X_{43} = m_3 \ddot{x}_Q \tag{D.16}$$

$$-Y_{32} + Y_{43} = m_3 \ddot{y}_Q \tag{D.17}$$

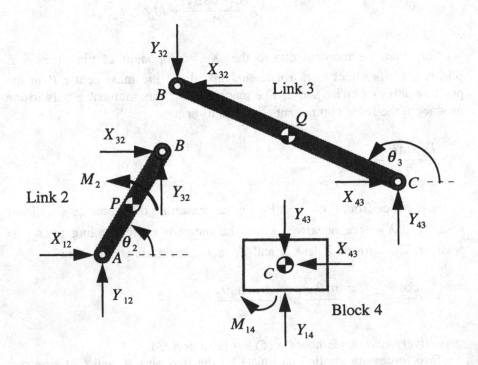

Figure D.9 Free body diagrams for the links

$$X_{32}\frac{L_3\sin\theta_3}{2} - Y_{32}\frac{L_3\cos\theta_3}{2} + X_{43}\frac{L_3\sin\theta_3}{2}$$

$$-Y_{43}\frac{L_3\cos\theta_3}{2} = I_3\ddot{\theta}_3.$$

(D.18)

Further explanation is needed regarding the signs of the moments appearing in (D.18). Inspection of the free body diagram for link 3 shows that for the position shown in Figure D.9 the moments associated with all the force components are positive. Note however that for the position of link 3 shown in that figure $\pi/2 < \theta_3 < \pi$. Hence $\sin\theta_3 > 0$ while $\cos\theta_3 < 0$, and all the terms in equation (D.18) are indeed positive for that configuration.

Two forces are applied to sliding block 4, the pin force F_{34} and the reaction with the ground, F_{14}. Since there is no friction, the ground applies normal force and $F_{14} = X_{14}$. In drawing the free body diagram for the block we have used Newton's third law

$$F_{34} = -F_{43}.$$

Therefore, the components X_{43} and Y_{43} appear in the negative sense instead of positive X_{34} and Y_{34}. In general, the sliding guide may apply also a moment that constrains the rotation of the sliding block. This moment is denoted in the diagram by M_{14}. Application of (3.8) to the sliding block 4 yields

$$-X_{43} = m_4\ddot{x}_C = m_4\ddot{r}, \qquad (D.19)$$

$$-Y_{43} + Y_{14} = m_4\ddot{y}_C = 0 \qquad (D.20)$$

and

$$M_{14} = I_4\ddot{\theta}_4 = 0 \qquad (D.21)$$

since the block translates along the x-axis with acceleration $\ddot{x}_C \equiv \ddot{r}$ and hence $\ddot{y}_C = \ddot{\theta}_4 = 0$.

At this stage the angular accelerations of links 2 and 3 and the linear acceleration of block 4 are known. Hence the nine equations (D-13) to (D.21) allow us in principle to determine the nine unknowns, namely: two components of each pin force A, B and C, the normal force and moment that the ground applies on the slider, and the external moment T_2. To actually solve the problem we still need to express the x- and y-components of the accelerations of P and Q in terms of the accelerations and angular accelerations of the links. It is easily shown that the acceleration of P, being the mid-point of link 2, is the average acceleration of points A and B. The various accelerations of the links are shown in Figure D.10. We attach cylindrical coordinates to point A of link 2 and find by (3.3c) that

$$\mathbf{a}_B = -\dot{\theta}_2^2 L_2 \mathbf{e}_r.$$

Noting that A is a fixed point, the x- and y-components of the acceleration of P are therefore

$$\ddot{x}_P = -\frac{1}{2}\dot{\theta}_2^2 L_2 \cos\theta_2 \tag{D.22}$$

$$\ddot{y}_P = -\frac{1}{2}\dot{\theta}_2^2 L_2 \sin\theta_2. \tag{D.23}$$

Similarly, the acceleration of Q is the average acceleration of points B and C. Hence by the acceleration diagram for link 3, shown in Figure D.10, we have

$$\ddot{x}_Q = \frac{1}{2}(\ddot{r} - \dot{\theta}_2^2 L_2 \cos\theta_2) \tag{D.24}$$

$$\ddot{y}_Q = \ddot{y}_P = -\frac{1}{2}\dot{\theta}_2^2 L_2 \sin\theta_2. \tag{D.25}$$

We can now show how to determine the required forces and moments. The moment M_{14} vanishes by virtue of (D.21). Knowing $\ddot{r}$ the force X_{43} is determined by (D.19). Then X_{32} can be found by (D.16), with $\ddot{x}_Q$ given by (D.24). Knowing X_{32} we can determine X_{12} from (D.13), with $\ddot{x}_P$ given by (D.22). We then solve equations (D.14), (D.15), (D.17) and (D.18) simultaneously

$$
\begin{bmatrix}
1 & 1 & 0 & 0 \\
-L_2\cos\theta_2/2 & L_2\cos\theta_2/2 & 0 & 1 \\
0 & -1 & 1 & 0 \\
0 & -L_3\cos\theta_3/2 & -L_3\cos\theta_3/2 & 0
\end{bmatrix}
\begin{pmatrix}
Y_{12} \\ Y_{32} \\ Y_{43} \\ M_2
\end{pmatrix}
$$

$$
= \begin{pmatrix}
m_2\ddot{y}_P \\
I_2\ddot{\theta}_2 + (X_{32} - X_{12})\,L_2\sin\theta_2/2 \\
m_3\ddot{y}_Q \\
I_3\ddot{\theta}_3 - (X_{32} + X_{43})\,L_3\sin\theta_3/2
\end{pmatrix}
$$

$$\tag{D.26}$$

and determine Y_{12}, Y_{32}, Y_{43} and M_2. Finally, from (D.20) we have $Y_{14} = Y_{43}$.

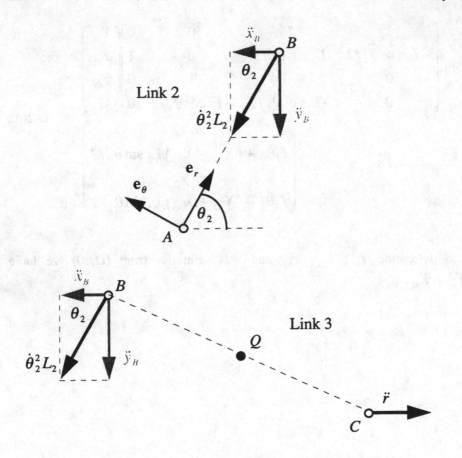

Figure D.10 Accelerations for the links

The magnitude of the pin force is given by

$$\left| F_{ij} \right| = \sqrt{X_{ij}^2 + Y_{ij}^2} \,, \tag{D.27}$$

for $i,j=1,2,3,4$. The external moment required to drive the mechanism with the prescribed motion is shown in Figure D.11, and the pin forces developed by the motion are given in Figure D.12. We see that the largest pin force applies on pin A at the position $\theta_2 = 0$.

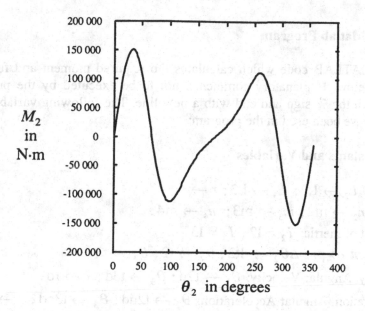

Figure D.11 The driving external moment

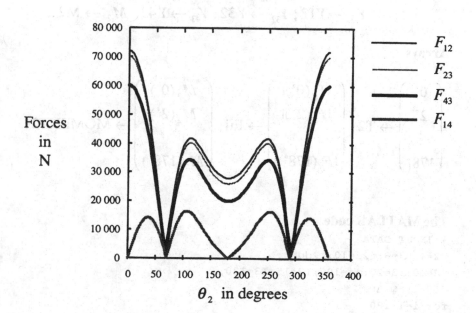

Figure D.12 Pin forces

D.6 Matlab Program

A MATLAB code which calculates the required moment and forces is given below. Explanatory comments, not to be executed by the program, start with the % sign and end with a new line. The following variables and arrays have been used in the program:

Constants and Variables

Length: $L_2 \rightarrow \text{L2}$; $L_3 \rightarrow \text{L3}$; $r \rightarrow \text{r}$.

Mass: $m_2 \rightarrow \text{m2}$; $m_3 \rightarrow \text{m3}$; $m_4 \rightarrow \text{m4}$.

Moment of inertia: $I_2 = \text{I2}$; $I_3 = \text{I3}$.

Angles: $\pi \rightarrow \text{pi}$; $\Delta\theta_2 \rightarrow \text{dt2}$; $\theta_2 \rightarrow \text{t2}$; $\theta_3 \rightarrow \text{t3}$.

Velocity, Angular Velocity: $\dot{\theta}_2 \rightarrow \text{t2d}$; $\dot{\theta}_3 \rightarrow \text{t3d}$; $\dot{r} \rightarrow \text{rd}$.

Acceleration, Angular Acceleration: $\ddot{\theta}_2 \rightarrow \text{t2dd}$; $\ddot{\theta}_3 \rightarrow \text{t3dd}$; $\ddot{r} \rightarrow \text{rdd}$;

$\qquad \ddot{x}_Q \rightarrow \text{xQdd}$; $\ddot{x}_P \rightarrow \text{xPdd}$; $\ddot{y}_Q \rightarrow \text{yQdd}$; $\ddot{y}_P \rightarrow \text{yPdd}$.

Force, Moment: $X_{43} \rightarrow \text{X43}$; $X_{32} \rightarrow \text{X32}$; $X_{12} \rightarrow \text{X12}$;

$\qquad Y_{12} \rightarrow \text{Y12}$; $Y_{32} \rightarrow \text{Y32}$; $Y_{43} \rightarrow \text{Y43}$; $M_2 \rightarrow \text{M2}$.

Arrays

$$\begin{pmatrix} 0° \\ 2° \\ \vdots \\ 178° \end{pmatrix} \rightarrow \text{T2}; \quad \begin{pmatrix} |F_{ij}(0°)| \\ |F_{ij}(2°)| \\ \vdots \\ |F_{ij}(178°)| \end{pmatrix} \rightarrow \text{Fij}; \quad \begin{pmatrix} M_2(0°) \\ M_2(2°) \\ \vdots \\ M_2(178°) \end{pmatrix} \rightarrow \text{MOMENT2}.$$

The MATLAB code

```
% INPUT DATA
L2=4;L3=8;t2d=10;t2dd=0;
m2=10;m3=20;m4=100;I2=15;I3=120;
dt2=2;  % Δθ₂=2°
for i=1:180,
      t2=(i-1)*dt2*pi/180;  % θ₂=0,π/90,...,179π/180
      % POSITION ANALYSIS
      t3=pi-asin(L2*sin(t2)/L3);  % Eq. (D.3)
      r=L2*cos(t2)-L3*cos(t3);  % Eq. (D.2)
```

```
% VELOCITY ANALYSIS
t3d=L2*t2d*cos(t2)/(L3*cos(t3));  % Eq. (D.7)
rd=L3*t3d*sin(t3)-L2*t2d*sin(t2);  % Eq. (D.8)
% ACCELERATION ANALYSIS
t3dd=(L3*t3d^2*sin(t3)-...
      L2*t2d^2*sin(t2))/(L3*cos(t3));  % Eq. (D.11)
rdd=L3*t3dd*sin(t3)+L3*t3d^2*cos(t3)-...
      L2*t2d^2*cos(t2);  % Eq. (D.12)
X43=-m4*rdd;  % Eq. (D.19)
xQdd=(rdd-t2d^2*L2*cos(t2))/2;  % Eq. (D.24)
X32=X43-m3*xQdd;  % Eq. (D.16)
xPdd=-t2d^2*L2*cos(t2)/2;  % Eq. (D.22)
X12=m2*xPdd-X32;  % Eq. (D.13)
yPdd=-t2d^2*L2*sin(t2)/2;  % Eq. (D.23)
yQdd=-t2d^2*L2*sin(t2)/2;  % Eq. (D.24)
A=[1 1 0 0;
   -L2*cos(t2)/2 L2*cos(t2)/2 0 1;
   0 -1 1 0;
   0 -L3*cos(t3)/2 -L3*cos(t3)/2 0];
b=[m2*yPdd;
   I2*t2dd+(X32-X12)*L2*sin(t2)/2;
   m3*yQdd;
   I3*t3dd-(X32+X43)*L3*sin(t3)/2];
X=inv(A)*b;  % Eq. (D.26)
Y12=X(1,1);Y32=X(2,1);Y43=X(3,1);M2=X(4,1);
% STORE DATA IN ARRAYS
T2(i,1)=t2*180/pi; F12(i,1)=sqrt(X12^2+Y12^2);
F32(i,1)=sqrt(X32^2+Y32^2);
F43(i,1)=sqrt(X43^2+Y43^2); F14(i,1)=abs(Y43);
MOMENT2(i,1)=M2;
   end
% PLOT RESULTS
plot(T2,MOMENT2); pause; plot(T2,[F12,F32,F43,F14]);
```

D.7 Problems

1. A spring of constant $k = 20000\,\text{N·m}$ and free length $L = 1\,\text{m}$ is attached between pin A and pin C of the slider-crank mechanism shown in Figure D.13. The links are uniform slender bars with mass per unit length $\rho = 10\,\text{kg/m}$, and the mass of the slider is 40 kg. Determine the moment M_2 which is required to drive the mechanism with constant

angular velocity of $\dot{\theta}_2 = 50$ rad / s, and the forces developed during the motion in the pins.

The moment of inertia about the centre of gravity of a slender bar with mass m and length a is given by

$$I = \frac{ma^2}{12}.$$

Neglect gravity, friction and the mass of the spring.

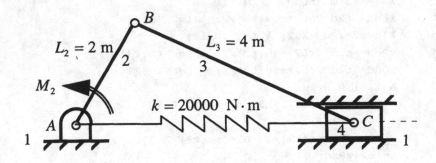

Figure D.13 Slider-crank mechanism with a spring

2. The links of the mechanism shown in Figure D.14 are uniform slender bars with mass per unit length $\rho = 15$ kg / m. Determine the moment M_2 required to accelerate the mechanism from $\theta_2 = 0$ rad / s to $\theta_2 = 50$ rad / s with constant acceleration $\dot{\theta}_2 = 5$ rad / s^2. What are the forces developed in the various pins?
Ignore gravity and friction. Use the following values

$$\overline{AB} = 2\,\text{m}, \quad \overline{BC} = 6\,\text{m}, \quad \overline{CD} = 4\,\text{m}, \quad \overline{PQ} = 4\,\text{m}, \quad \overline{QR} = 4\,\text{m},$$
$$\overline{BP} = 3\,\text{m}.$$

3. What is the moment M_2 that is needed to drive the mechanism of Figure D.15 with constant angular velocity $\dot{\theta}_2 = 40$ rad / s? When operating under this condition, what are the forces developed in the pins?
Ignore gravity and friction, and use the following data:

$$\overline{AB} = 1\,\text{m}, \ \overline{AC} = 2\,\text{m}, \ \overline{BD} = 4\,\text{m}, \ m_3 = 30\,\text{kg}, \ I_3 = 7\,\text{kg}\cdot\text{m}^2.$$

Links 2 and 4 are slender bars with $m_2 = 10 \text{ kg}$ and $m_4 = 20 \text{ kg}$.

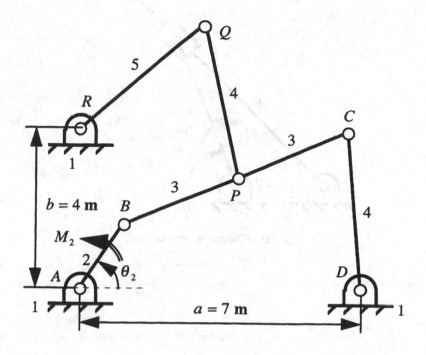

Figure D.14. Multi-bar linkage

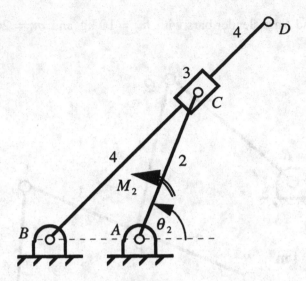

Figure D.15 Slider-bar mechanism

INDEX

MATHEMATICAL MODELLING:
Theory and Applications

---

1. M. Křížek and P. Neittaanmäki: *Mathematical and Numerical Modelling in Electrical Engineering*. Theory and Applications. 1996
   ISBN 0-7923-4249-6

2. M.A. van Wyk and W.-H. Steeb: *Chaos in Electronics*. 1997
   ISBN 0-7923-4576-2

3. A. Halanay and J. Samuel: *Differential Equations, Discrete Systems and Control*. Economic Models. 1997 ISBN 0-7923-4675-0

4. N. Meskens and M. Roubens (eds.): *Advances in Decision Analysis*. 1999
   ISBN 0-7923-5563-6

5. R.J.M.M. Does, K.C.B. Roes and A. Trip: *Statistical Process Control in Industry*. Implementation and Assurance of SPC. 1999
   ISBN 0-7923-5570-9

6. J. Caldwell and Y.M. Ram: *Mathematical Modelling*. Concepts and Case Studies. 1999 ISBN 0-7923-5820-1

KLUWER ACADEMIC PUBLISHERS – DORDRECHT / BOSTON / LONDON